CW00787618

AIRFIX
magazine guide 10

Luftwaffe Camouflage of World War 2

Bryan Philpott

Patrick Stephens Ltd
in association with Airfix Products Ltd

First published – September 1975

ISBN 0 85059 213 5

Cover design by Ian Heath

Drawings by Martin Holbrook

Text set in 8 on 9 pt Helvetica Medium by Blackfriars Press Limited, Leicester.
Printed on Fineblade cartridge 90 gsm and bound by The Garden City Press, Letchworth, Herts.
Published by Patrick Stephens Limited, Bar Hill, Cambridge CB3 8EL, in association with Airfix Products Limited, London SW18.

Contents

Acknowledgments

I am indebted to the following whose enthusiastic and unselfish help has made my task easier than it might have been.

Martin C. Windrow for photographs and permission to quote from his published works; Martin Holbrook whose fine drawings are greatly appreciated; Carl J. Weigand for help in translating German letters; the editor of *IPMS Magazine* and the executive committee of that organisation for giving permission to draw on their publications; and many other modellers too numerous to mention individually, who have contributed both large and small items; and finally my wife Susan who spent many hours mixing paints in an attempt to match various colour chips in my possession.

Editor's introduction

The markings and camouflage styles of aircraft used by the German Luftwaffe in World War 2 has always proved a popular subject among model makers, and the range of kits available of aircraft of this period are testimony to the seemingly endless fascination in which the period is held.

This book is by no means a definitive work on the subject, and should not be regarded as such: indeed, a definitive book would, in any case, be impossible to write today, 30 years after the end of the war, since so many records have been lost, peoples' memories are short and not always reliable, and colour photographs of the period rare. The intention here has been instead to draw the broad outlines of the subject for the modeller who wishes to centre his hobby on the Luftwaffe but has yet to gain sufficient experience to delve more deeply into the matter.

In this book, Bryan Philpott, a long-standing contributer to *Airfix Magazine* and editor of *IPMS Magazine,* describes the basic schemes and colours used to camouflage, identify and decorate German aircraft used in World War 2, in an attempt to help the newcomer to the hobby as well as sow the seeds from which a more detailed study may grow.

However, even experienced modellers will find much here of interest, and no other outline guide to the subject yet exists in such a readily assimilable — and cheap — form. For those who wish to delve more deeply into the subject of German markings, the four-volume set of books by Karl Ries, published by Verlag Dieter Hoffman, referred to later, is a 'must', together with the Kookaburra publication *Luftwaffe Camouflage and Markings,* by K. Merrick (of which only the first volume is so far available).

To begin with, though, this book gives an outline of the development of German aircraft camouflage and marking schemes, from World War 1 through the inter-war period, the revelation of the Luftwaffe as a fighting force after Hitler and the Nazis came to power, operations in the Spanish Civil War and, of course, during World War 2.

Light and heavy fighters, night fighters, ground attack aircraft, bombers, reconnaissance, transport, maritime and training types all fall within the book's scope, and there is also a useful appendix showing how to mix accurate camouflage colours using Airfix enamel paints, as well as a chapter describing briefly the best ways of applying camouflage to plastic model aircraft kits.

Luftwaffe organisation and individual aircraft identification, national markings, individual kill markings and squadron and personal insignia are all covered, making this book *the* ideal introduction to the subject.

BRUCE QUARRIE

one

Organisation and basic schemes

The evolution of the aeroplane as a weapon for waging war moved very rapidly from infancy to maturity during the four years of World War 1. At the start of the Great War the generals regarded the frail string and canvas creations that had started to invade the skies, as a passing invention that mainly frightened cavalry horses but had limited military use. Those who thought otherwise could initially only visualise them as useful artillery spotters or perhaps as instruments for gathering information on enemy ground movements. But when some intrepid crews started carrying weapons to fire at each other and then took this form of offensive one step further by adding guns to the aircraft, the first faltering steps in turning the aeroplane into a powerful weapon in any nation's armoury had been taken.

It was not long before aircraft were designed to perform specific tasks such as air-to-air fighting, bombing and reconnaissance, and parallel to this came the specialised use of weapons. By the end of the war it was apparent that he who controlled the air also controlled the ground, and further developments in the following years simply extended the tactical and strategic use of air power.

Early fliers tended to regard their machines as cavalry of the air, and there was a sense of chivalry between opposing airmen that can be likened to that which existed between knights of a former era. Like their forefathers, the new knights of the air had a strong leaning towards colourful decoration and bedecked their aircraft with colour schemes and badges that did little for concealment but a lot for morale and easy identification. The absence of radio equipment meant that all communication had to be visual by hand signals, or in the height of combat, by easily recognised symbols, colours, badges or streamers.

Both the Royal Flying Corps and the German Air Force made rudimentary attempts at camouflage but tended to concentrate their efforts on aircraft that were used for low-level reconnaissance or concealment on the ground. In other words, to make identity from above difficult. This is not to say that serious thought had not been given to the art of camouflage, but it was to be years before the exacting science of concealment became a well-practised art.

The Royal Flying Corps appears to have made much more serious attempts in the use of drab colours that matched environmental conditions, whereas the German Air Force were generally much more flamboyant. Few readers will not have heard, or read about, the exploits of Baron Manfred Von Richtofen and his famous Flying Circus, in which all aircraft were painted colours that were primarily aimed at being seen at extreme distances. The flamboyancy of the German fliers started a tradition

This Ju 86 shows the early splinter of light and dark green. The Swastika is painted on a white disc and dates the aircraft as being photographed in late 1938 or early 1939. The splinter camouflage can be clearly seen on the wing of the photographic aircraft (Hans Obert via Martin Windrow).

that was to continue, to a much lesser degree, during World War 2, but before that a period of depression, followed by a secret build-up of the new German Air Force was to occur.

The cessation of hostilities in November 1918 saw the end of the Imperial German Air Force and the Treaty of Versailles, signed in 1919, forbade the re-establishment of a military air arm in Germany, although it did not curtail the manufacture of civil aircraft. Some of the aircraft used during the war were modified for civil use and a simple numbering system was devised in which the individual aircraft number was preceded by the letter 'D', indicating 'Deutschland'. Thus the first civil registration system to be used on German aircraft came about.

The aircraft carried consecutive numbers and the first to carry such a registration was the prototype low-wing cantilever monoplane, the Junkers F 13, which received the registration D-1. No attempt was made to allocate aircraft of the same classification individual blocks or groups, each type receiving a consecutive number as it was registered.

Geschwader Stab

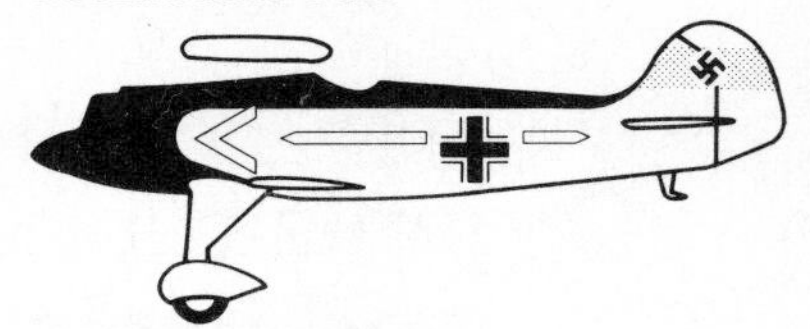

Geschwader Kommodore

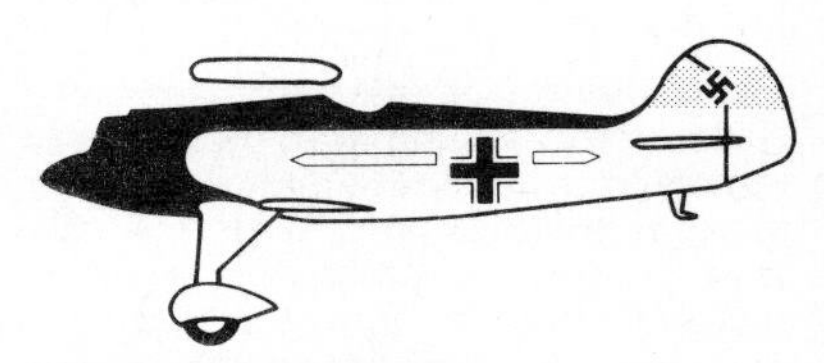

Geschwader Adjudant

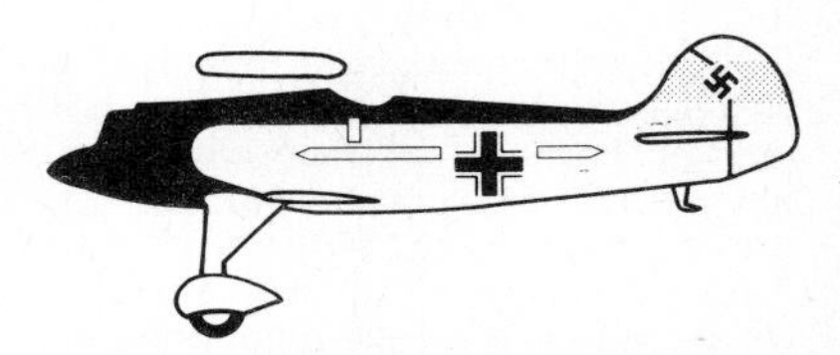

Geschwader Technical Officer

Organisations and markings of Geschwader pre-1938

The markings used to identify Geschwader and Gruppen aircraft prior to 1938 were similar in many respects to those carried after this date and shown on pages 24-5.

Geschwader and Gruppen Staff aircraft carried similar identification marks on their top wing surfaces. For example, the aircraft flown by the Geschwader Kommodore carried the chevron at each wing tip, inboard of the cross, with a pointed line going into the point of each chevron over the whole span. Gruppen aircraft carried the chevron style marking on the centre section but did not have the individual Gruppe identifying symbol, so an aircraft carrying the chevron of the Gruppe Kommanduer could not be identified from above as belonging to any particular group.

The individual Gruppen identifying symbols were carried behind the chevron but in front of the fuselage cross, so the I Gruppe Kommanduer's aircraft had no marking behind the chevron, the II Gruppe Kommanduer's had a horizontal bar and the III Gruppe a horizontal wavy line.

Staffeln aircraft carried their individual identifying number in the staffel colour as well as a symbol showing the staffel. So aircraft of 1, 4 and 7 Staffeln had white markings with no other identifying symbol, those of 2, 5 and 8 had a vertical bar around the cowling and tail and those of 3, 6 and 9 had a white disc in the two positions. An aircraft carrying a white number followed by a horizontal bar would have been a 4 Staffel aircraft of II Gruppe. Similarly one with a horizontal wavy bar, plus a disc on the front cowling and tail would have been a machine of the 9 Staffel of III Gruppe.

Organisation and markings of Geschwader pre-1938

Gruppe Stab

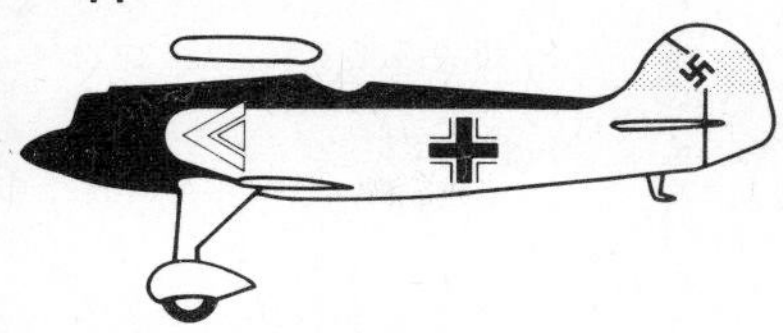

I Gruppe Kommanduer

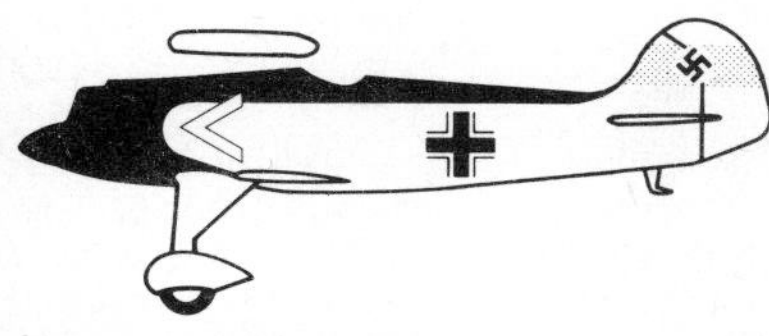

I Gruppe Adjudant

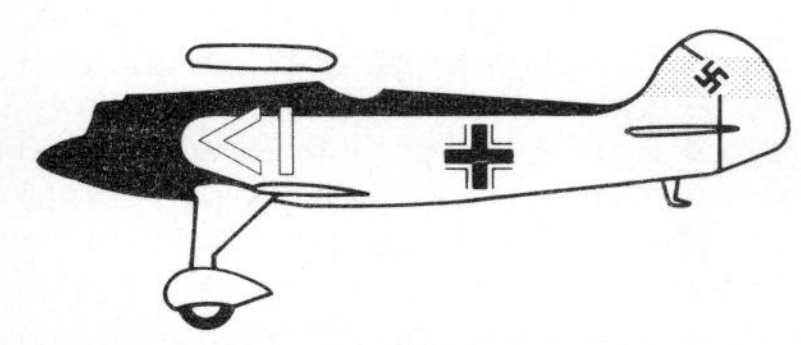

I Gruppe Technical Officer

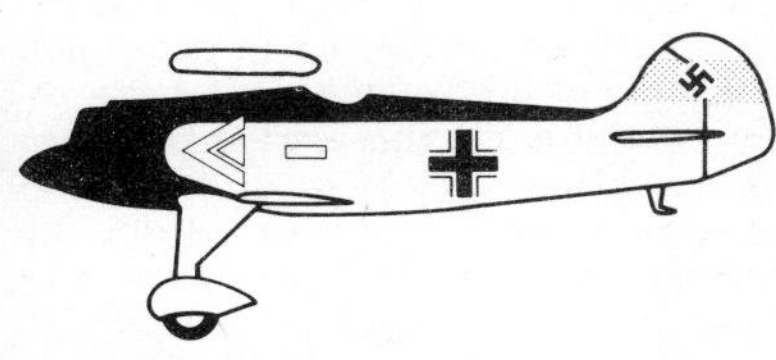

II Gruppe Kommanduer

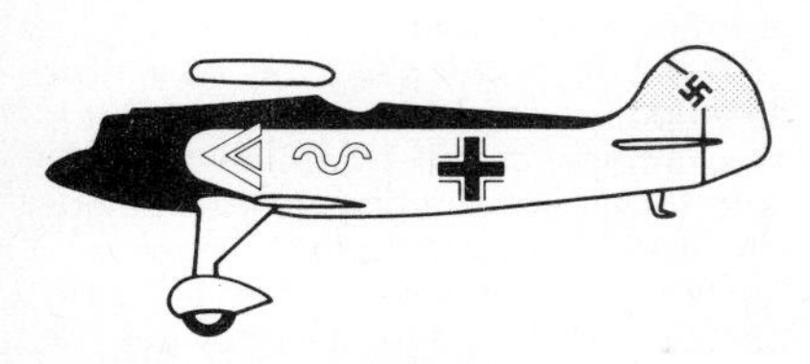

III Gruppe Kommanduer

Staffel

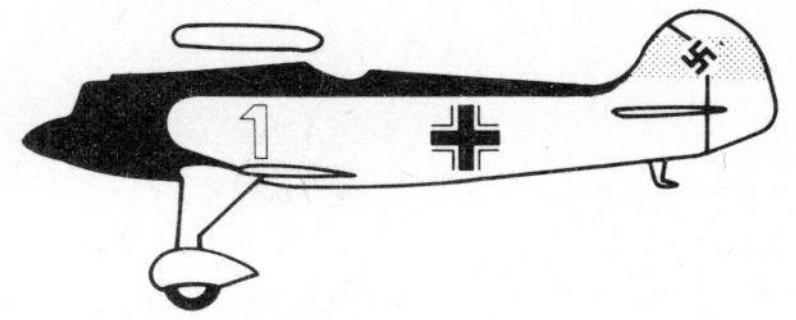

1, 4 and 7 Staffeln

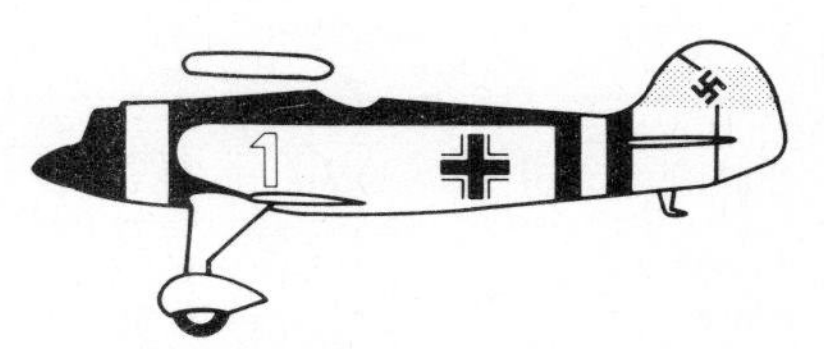

2, 5 and 8 Staffeln

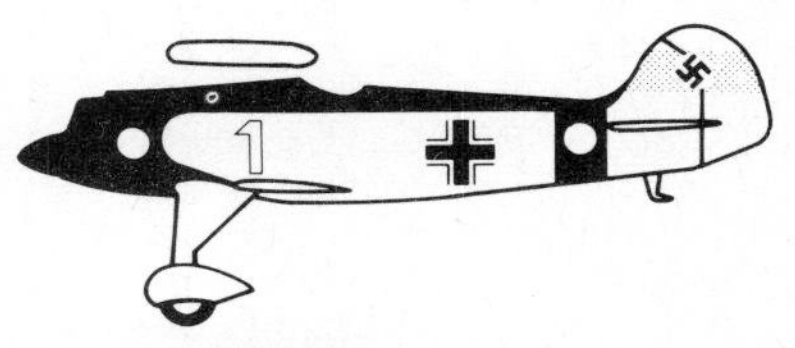

3, 6 and 9 Staffeln

This system lasted for approximately 15 years during which some 3,000 numbers were issued, but by 1935 most civil aircraft carried a new code consisting of four letters prefixed by 'D' which was introduced in 1933 when the National Socialist Party came to power.

The new registration system allowed for the division of aircraft into specific categories, each based on a variety of parameters which included the weight, number of crew and quantity of engines, as well as differentiating between land and seaplanes.

The class to which the aircraft belonged was indicated by the first letter of the four-letter group which followed immediately after the national code letter 'D'. The remaining three letters identified the particular aeroplane and were allocated in strict sequence. The complete breakdown of this coding system is as follows:

Class	Registration group	Crew	Loaded weight	Engines
A1	D-YAAA to D-YZZZ	1	up to 500 kg	1
A2	D-EAAA to D-EZZZ	1 to 3	up to 1,000 kg	1 or 2
B1	D-IAAA to D-IZZZ	1 to 4	up to 2,500 kg	1 or 2
B2	D-OAAA to D-OZZZ	1 to 8	up to 5,000 kg	1 or 2
C	D-UAAA to D-UZZZ	over 6	over 2,500 kg	1
	D-AAAA to D-AZZZ	over 6	over 2,500 kg	multi
Seaplanes				
A1	D-YAAA to D-YZZZ	1	up to 600 kg	1
A2	D-EAAA to D-EZZZ	1 to 3	up to 2,000 kg	1
B	D-IAAA to D-IZZZ	1 to 4	up to 3,500 kg	1
B	D-UAAA to D-UZZZ	1 to6+	up to 3,500 kg	1
C	D-AAAA to D-AZZZ	over 6	over 3,500 kg	multi

During the clandestine build-up of the new German Air Force, many so-called civil developments were little more than thinly disguised military types, and these, together with similarly destined high performance and experimental machines, were usually coded within the D-IAAA to D-IZZZ series.

The original Alpha/Numerical code instituted in 1918 was usually marked on the fuselage sides of the aircraft concerned, but it was occasionally also painted on the rudder. In some cases the 'D' prefix was not used but by 1920 an effort had been made at standardisation which saw the registration marked on both sides of the fuselage, located between the tailplane and wing location, the upper and lower surfaces of port and starboard mainplanes on monoplanes, and on both port and starboard lower surfaces of the bottom wings on biplanes as well as the port and starboard surfaces of the top wings.

The positions for the four-letter codes was the same as for the Alpha/Numerical registrations and the letters were applied following a specifically laid down spacing and geometric pattern detailed in the accompanying drawing.

Prior to the emergence of the National Socialist Party, aircraft had carried no other national insignia, but the Party introduced this in the form of markings on the vertical tail surfaces. These markings consisted of the Party flag which was a black *Hakenkreuz* (Swastika) on a white disc superimposed on a red background. This was initially confined to the rudder only on the port side, but soon spread across both the fin and rudder. The starboard side carried three equal bands, occupying the same area as the port side marking, comprising the German national colours of black, white and red. This style of marking lasted from 1933 until 1935 when the existence of the Luftwaffe was revealed to a startled world, at which time the port side marking of the Party flag was also painted on the starboard side to replace the three broad bands.

Civil registrations were retained by the Luftwaffe until early 1936 when distinct changes started to occur and among the first of these was the reintroduction of the *Balkenkreuz* or Greek cross, as used in the latter stages of World War 1, to identify the aircraft as a military machine. This type of cross, of which there were to be six major types before the end of World War 2, was applied to the upper and lower surfaces of the wings and both fuselage sides.

The civil codes gave way to a new five-figure/letter combination which was applied to all front-line aircraft, thus leaving do doubt as to the military purpose of the machine. There was a laid down order for the military code system and this saw the first pair of numerals positioned before the cross on the fuselage, followed by a single letter and two more numerals. The coding was repeated on the upper sur-

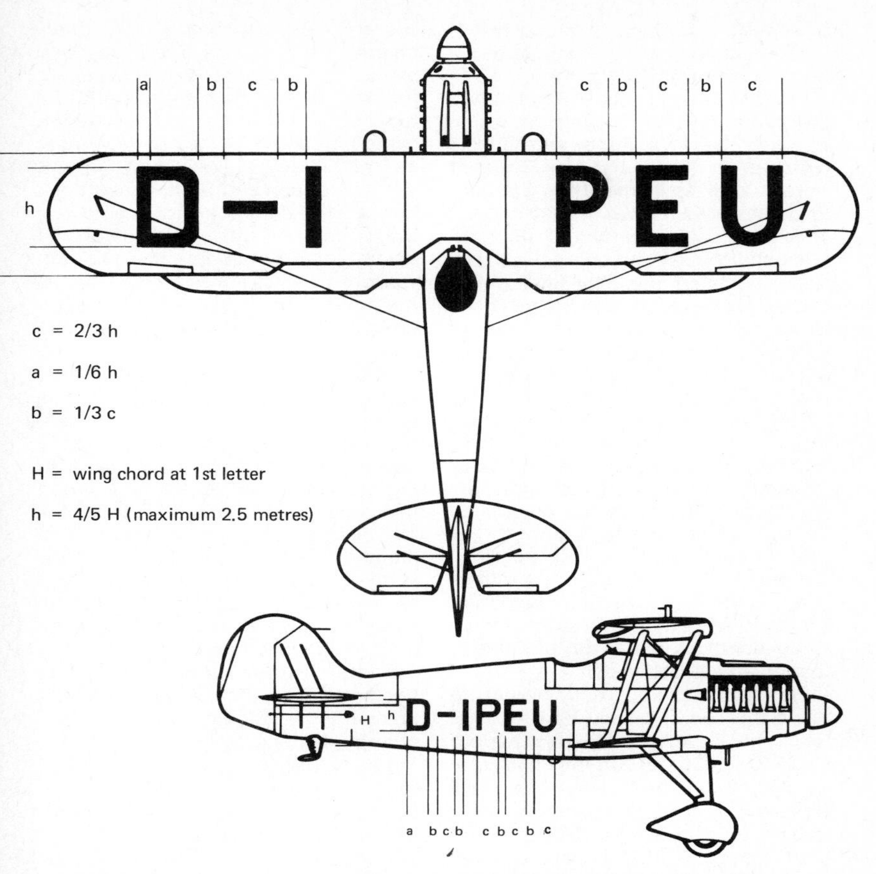

face of the top wings and lower surface of the bottom wings but positioned so that the first pair were inboard of the wing cross, a single letter on the centre section, and the remaining two inboard of the starboard cross. In cases where the central letter interferred with undercarriage struts, etc, on the under surfaces, this was grouped with the last pair under the port wing.

The organisation of the Luftwaffe at this time was such that the five-character code was determined by a rather complex formula based on the disposition and strength of the units concerned. The strength was divided between six Local Air Commands, called *Luftkrieskommandos,* each of which was based on a particular area in which was situated their headquarters. The first character in the code designated the Luftkreiskommando to which the aircraft was assigned, the location of these being as follows:

Luftkreiskommando I:	Königsberg
Luftkreiskommando II:	Berlin
Luftkreiskommando III:	Dresden

Position and spacing of pre-war codes

The five letter civil code used during the clandestine build-up of the Luftwaffe was painted on the fuselage sides and the top and lower wing surfaces spaced as shown.

Luftkreiskommando IV: Munster
Luftkreiskommando V: Műnchen
Luftkreiskommando VI: Kiel

Within each Luftkreiskommando there were several *Geschwader*, which in turn were divided into three *Gruppen* in each, and the Gruppen were further split into three *Staffeln*. Each Geschwader was formed in strict chronological order within its Luftkreis and the second character in the code indicated this. These first two characters were followed by the Greek Cross, which at this time was of symmetrical design with the sides of the arms outlined in white which in turn had a thin black trim.

The first character after the cross indicated the individual aircraft number within the Staffel and this was followed by a number that indicated the Gruppe. The code was completed by a fifth number which showed the Staffel concerned. It will be seen therefore, that although the codes enabled the Local Air Command, and Staffel to which the aircraft belonged, to be identified, it was not possible to identify the Geschwader.

This becomes apparent by breaking down the codes carried by a Heinkel He 51 of JG 132 during this period: Code 21 + A25. 2=Luftkreis II, Berlin; 1=JG 132 which was the first Geschwader formed in Luftkreis II; + Balkenkreuz or Greek Cross; A=aircraft letter; 2=the second Gruppe; 5=the fifth Staffel.

So, although from the codes the aircraft could be identified as being 'A' of the fifth Staffel of the second Gruppe, belonging to the first Geschwader formed in Luftkreis II, the fact that the Geschwader is JG 132 is not revealed.

For ease of identification in written documentation the Gruppen was written in Roman numerals while the Staffeln used Arabic.

An interesting deviation to the five-character code was the style used on seaplanes which came under the control of Luftkreis VI. Since there were insufficient aircraft to warrant a full Geschwader, the second code character was not needed and was replaced by an '0', thus during the immediate pre-war period while the five-character code was used, all seaplanes carried codes commencing 60, followed by the Balkenkreuz and the remaining three digits which gave the same indication as they did for other aircraft, that is, individual aircraft letter, Gruppen and Staffeln numbers. It was also common practice to use the '0' code on other types of aircraft where their prime function did not warrant a full Geschwader such as the reconnaissance aircraft within Luftkreiskommando II which carried the codes 20.

In addition to the five-character code detailed, various fighter units also carried colours as an aid to identification. These colours usually covered the whole of the engine cowling and were often swept back along the spine of the fuselage to the tail fin, but also appeared as a broad band painted around the fuselage just forward of the empenage. The application tended to vary between units but the coloured cowling was usually common to all types operated.

Some fighter units had titles bestowed upon them, thus JG 132 was known as the Richtofen Geschwader after the famous World War 1 ace, and naturally used red — after the Red Baron — as its identifying colour. Colours allocated to the Jagdgeschwader (Fighter Wings) were: JG 131 (based at Jesau): Black; JG 132 (based at Dőberitz): Red (Richtofen); JG 134 (based at Dortmund): Brown (Horst Wessel); JG 232 (based at Bernburg): Green; JG 233 (based at Bad Aibling): Blue; JG 234 (based at Cologne): Orange.

In addition to the identifying codes and national insignia mentioned, it was laid down that all aircraft had to display in an easily accessible position on the fuselage, wings, ailerons, and flaps, the name and address of the manufacturer. The engine/s also had to carry a metal plate detailing the name and address of the manufacturer, type, designation, series, works number and year of construction, the horse power developed and the maximum permissible RPM. This detail is of little use to modellers who confine their activities to small scales, but it does enable those who favour larger models to add a little extra detail that adds to the authenticity of the model.

The manufacturers' plates just mentioned were not often visible and cannot therefore be considered as part of the external finish of German aircraft, but one legend that should be added to any representative model of the period is that which was painted on the port side of the fuselage, usually at the rear below the tailplane. This was a stencilled table in letters 25 mm high and 4 mm thick giving the following information: Name and address of owner; Empty weight, payload and maximum permissible all-up-weight in kilograms; Maximum permissible number of persons, including crew, to be carried; and Date of the last Major Inspection and the date of the next Major Inspection.

During the years when the Luftwaffe was being rebuilt, the owner's name and address often referred to a Flying School or Club, and this continued after the adoption of the National Socialist Party Hakenkreuz marking on the fin/rudder. Naturally the extremely complicated identification system soon proved unwieldy and quickly defeated its object of providing quick recognition, since it relied on pilots being able to remember the meaning of the five-character code. So it came as no real surprise when, in July 1936, a new directive was promulgated in which the traditional colours were retained but a simpler system of markings replaced the five-character code.

The new system meant that each Staffel aircraft would carry a number between one and 12, painted in white with a black outline, on each side of the nose and the uppper and lower wing centre-sections. Identification of the Staffel within the Gruppe would be achieved by additional markings on the coloured areas, these being: Staffeln 1, 4, and 7 — No markings; Staffeln 2, 5, and 8 — A white band painted around the cowling and superimposed on the broad coloured band around the rear fuselage; Staffeln 3, 6, and 9 — A white disc on the cowling which was also repeated on the fuselage band.

The Gruppe within the Geschwader also carried an identifying symbol painted on the fuselage side aft of the number but before the Balkenkreuz, these symbols being: I Gruppe — No marking; II Gruppe — A horizontal bar; III Gruppe — A horizontal wavy line.

Aircraft operated and flown by the Staff officers of the Geschwader and Gruppe, known as *Geschwaderstab* and *Gruppenstab,* carried identifying symbols instead of the numbers carried by Staffel aircraft. These symbols consisted of a combination of chevrons and bars and were painted white outlined in black and were as illustrated. Staffeln commanders usually flew aircraft carrying the number 1 but were not identified in any other way.

Official orders stated that the new identification codes and symbols should be applied to all Luftwaffe aircraft by September 1 1936.

Prior to the first attempts at coordinating a camouflage system, most Luftwaffe aircraft had been painted silver or a greenish-grey colour with the codes and national insignia in black. The only other colours used in any quantity were the identifying colours for various fighter units, the white with black outlined Staffel and Geschwaderstab/Gruppenstab codes and symbols, plus the red background to the Hakenkreuz carried on the fin/rudder.

Before going on to take a closer look at the development and use of various camouflage schemes used by the Luftwaffe during World War 2, it is worth outlining the markings used by the aircraft sent to Spain as part of Germany's military aid to the Spanish Nationalist forces who were engaged in an attempt to overthrow the Republican Government. By late 1936 the Germans had committed approximately 200 aircraft to this conflict and they became known as The Condor Legion.

two

The Condor Legion

From 1936 to the early part of World War 2, German aircraft started to appear in a variety of colour schemes which reflected the German Air Ministry's *(Reichsluftfahrtministerium,* hereafter abbreviated to RLM) efforts to establish a method of camouflage for military aircraft. During this period the RLM worked in close co-operation with aircraft manufacturers and consultants, in an attempt to compile specific camouflage systems using a laid-down range of colours developed solely for this purpose.

With the thoroughness that underlined all aspects of the work of the RLM, a schedule was laid down in which every paint group was identified by a two-figure number, individual aircraft paints by a four-figure code, and the actual colour pigment by a further two-figure code. The latter started at 00 and eventually reached 82, these colours being defined in Appendix 2.

During 1937 the most common scheme to be seen on fighters was light grey on all upper surfaces and pale blue on the undersides, but by early 1938 the light grey started to be replaced by *Schwarzgrün,* which was a matt dark green, the literal translation being 'black-green', and this corresponded to colour 70.

Colours and camouflage schemes to be used on individual types of aircraft were specified in the Technical Manuals for each type and generally speaking these were strictly adhered to, although in later years, as with all camouflage schemes employed by any air force, these were often modified both in colour and design as local conditions and availability of supplies dictated. But in 1936, when the aircraft of the new Luftwaffe first tasted combat in the Spanish skies, the embryo camouflage schemes were painted in strict accordance with directives.

It was also during this period that the last major change to occur as far as Gruppe and Staffel markings was introduced, although these markings were not used on the aircraft operating with the Spanish Nationalist forces. These changes will be detailed in Chapter 3 as they apply more to the 1939-45 period than they do to the Condor Legion.

The first aircraft to be sent to Spain were 20 Ju 52/3ms to be used in the transport or bomber roles, and accompanying them were six He 51

A Bf 109B-2 of the Condor Legion in Spain. The aircraft is finished light grey overall. The '6' indicates it is a Bf 109 while the 53 is the aircraft's individual code number (Hans Obert via Martin Windrow).

He 51 of 4/J88 Legion Condor Spain 1939, it carries the three tone grey green brown pre-war scheme with light blue undersides. The wing tips and spinner are white (Hans Obert via Martin Windrow).

fighters, all these machines being finished in the then standard light grey finish. By May 1937 the force had grown to about 200 aircraft and it was these that became a self-governing air force using the title 'Legion Condor'.

The elements of the force consisted of a bomber group — *Kampfgruppe* — of three flights of Ju 52s, this being designated Kampfgruppe 88 (K/88); a fighter group — *Jagdgruppe* — of four flights of He 51s carrying the nomenclature Jagdgruppe 88 (J/88); a quantity of 24 He 70 and He 45 aircraft forming a reconnaissance element designated *Auflarungsstaffel* 88 (A/88); and a force of He 59 seaplanes designated *See-Aufklärungsstaffel* 88 (AS/88).

The inclusion of some form of national markings to identify aircraft of the combatants was, of course, a necessity, but the colours of the Nationalists (red/yellow/red) were too close to those of the Republicans (red/yellow/purple), so a marking with greater contrast was introduced.

As the Condor Legion were fighting with the Nationalists they had to adopt their markings, so the aircraft concerned had their rudders painted white and had a black diagonal cross superimposed on them. Wing and fuselage markings were a simple black disc, that on the fuselage being spaced between the identifying numerals which were allocated to individual aircraft as follows: Dornier Do 17 — 27; Fiesler Fi 156 — 46; Heinkel He 51 — 2; He 45 — 15; He 59 — 71; He 70 — 14; He 111 — 25; Henschel Hs 123 — 24; Hs 126 — 19; Junkers Ju 52 — 22; Ju 86 — 26; Ju 87 — 29; W 34 — 43; Messerschmitt Bf 108 — 44; and Bf 109 — 6.

The above code numbers were painted on the left of the black fuselage disc on both sides, so viewed from the port side they appeared towards the nose, and from the starboard towards the tail. To the right of the disc was painted the aircraft's individual code number which, for security reasons, usually had a random starting point. There were, of course, exceptions to the rule and on some occasions aircraft could be seen with both the identifying type number and individual code applied to one side of the disc. When this happened the type code and number were separated by a hyphen.

The fin and rudder of this He 111B-2 of KG88 shows the typical rudder marking for Condor Legion aircraft as well as a forerunner of aircraft decorative art in the form of the crews' pet Scottie dog Peter which flew on operations with them. Full code of the aircraft is 25 ● 15 (Hans Obert via Martin Windrow).

Messerschmitt Bf 109E-1

This Bf 109E-1 is in the markings of 2/J88 as seen in the Spanish Civil War. The aircraft has light grey (63) upper surfaces and light blue (65) on the undersurfaces. Fuselage insignia is the solid black disc which sometimes had a white cross in the centre. The rudder is white with a diagonal black cross. The spinner is white as are the undersides of the wingtips. The '6' is the identifying code for Bf 109 aircraft while the 128 is the aircraft's individual identifying code. The top hat is the unit emblem which appeared in a number of forms during World War 2. It was also the basis for many pilot's personal insignia and featured in one way or another in several unit badges. The aircraft shown has the wing roots and areas around the exhausts painted black in a form that could be seen on many aircraft of the Legion Condor.

The initial system was modified rather quickly after the start of hostilities and one of the early changes was the addition of an extra black disc painted on the nose of some aircraft, particularly the He 59 where the bulk of the engines and wings could obscure the normally placed fuselage marking. The most significant change, however, was the addition of a white St Andrew's cross painted across the full diameter of the top and undersurface black discs as well as overpainting the wing tips white. The St Andrew's cross was also used on some types without the black disc on the top surfaces of wings, when it usually covered the full chord.

Although there were variations of all types and styles of markings which makes a generalisation impossible, it is fair to claim that the monoplane fighters, such as the Bf 108 and Bf 109, more often than not followed a more standard pattern than their earlier contemporaries. This pattern being the black fuselage disc with the wing surfaces having a similar disc overpainted with the diagonal cross.

National markings aside, the basic camouflage schemes to be seen in Spain generally followed the same patterns as those that were being introduced into the Luftwaffe. Most aircraft delivered during the initial stages of the conflict were painted light-grey (63) overall but it was not long before overpainting using a variety of other colours started to be introduced.

The He 51s of J88 had their undersurfaces painted light blue (65) and some of these aircraft had a disruptive pattern of brown (61) and green (62) painted over the light grey top surfaces. This pattern followed the angular segmented pattern as defined for bombers, which was to be more commonly seen on the He 111s and Do 17s.

Light grey (63) and light blue (65) appear to have been fairly common for the Bf 109 fighters, at least when they first appeared, but later many of them carried overall green (62) or brown (61) top surfaces to make them less visible when operating over the Spanish terrain. Examples of Bf 109s carrying green on the fuselage and brown on the top surfaces of the wings and tailplanes have been recorded, proving once again that very thorough research is needed by the modeller who wishes to specialise in German aircraft used in this particular field of operations.

The arduous extremes of weather quickly took a toll of the paintwork of all aircraft, and field work was often carried out with the only paint available at the location concerned. This often took the form of the use of RLM Grau (02) which was a standard primer finish and was sometimes used to cover large areas or even overall finishes.

Spanish operations also saw a gradual introduction of personal and unit emblems, and some of the Staffel badges employed at this time carried on well into World War 2. Some of the more commonly known examples were the diving raven of I/J88; the top hat insignia of 2/J88; and the Condor carrying a bomb between its claws of 1/K88. The positioning of such insignia varied and could be seen painted below cockpits, just forward of fuselage codes, on top of the black national insignia disc and even on the wheel spats of some Ju 87s.

These brief notes on the Condor Legion are not intended to be exhaustive but have been included simply to illustrate how the early development of Luftwaffe camouflage first saw the light of day during the Spanish Civil War. Many of the schemes used reflected those being tried in Europe and most of them continued to a lesser or greater degree during the early years of World War 2.

The following sample schemes for aircraft used by the Condor Legion are intended as a guide to a cross section of the ones in use, and serve to illustrate the variations that abounded at this time.

Heinkel He 51B of 4/J88: Undersurfaces — Hellblau (65); Top surfaces — Segment camouflage of Grün (62), Dunkelbraun (61) and Hellgrau (63); Spinner and rudder — White (21); Codes — 2 ● 111; Markings — Black fuselage discs with Staffel badge of ace of spades on white diamond painted over it. Black diagonal cross superimposed over rudder covering full chord and height. White full chord St Andrew's crosses on top wings.

Bf 109E of 2/J88: Undersurfaces — Hellblau (65); Top surfaces — Hellgrau (63); Spinner and rudder — White (21); Codes — 6 ● 136; Markings — Black fuselage discs with Staffel badge of top hat forward of aircraft code 6. Black diagonal cross superimposed over rudder covering full chord and height. White St Andrew's crosses superimposed over black discs on top and bottom surfaces of wings. This aircraft also had exhaust area painted black which was styled back over wing roots to wing trailing edges.

Heinkel He 111B-1 of 1/K88: Undersurfaces — Hellgrau (63); Top surfaces — Hellgrau (63); Spinner — Hellgrau (63) for rear portion, White (21) front portion; Rudder — White (21); Codes — 25 ● 3; Markings — Black diagonal cross superimposed over rudder as for previous examples. Black fuselage discs with diving Condor badge of the Staffel superimposed over it. Top and bottom surfaces of wings, black discs with St Andrew's crosses in white occupying full diameter. This aircraft also had name 'PEDRO' with figure 1 under the 'D' painted in white on port side of nose just aft of front gunner's window. This name was common to the He 111 while many of the Do 17s carried the appelation 'PABLO'.

three

Fighters 1939–1945

The first major colour to be used on Luftwaffe aircraft during the transition period to full camouflage systems was Schwarzgrün (70), and most of the early Bf 109Es appeared painted in this colour on all their top surfaces and fuselage sides, the undersurface colour being Hellblau (65), but in 1939 the RLM issued a colour chart listing 21 colours that were to be used on various parts of aircraft and in a variety of layouts. By 1941 three additional colours had been added and later the same year a third version was issued which introduced four more colours.

It was during the spring of 1939 that the new system of identifying codes, first mooted the year before, started to appear on aircraft, superseding all previous instructions and staying in force until the end of the war. This was a much more simple method of numbering and provided the necessary distinction for individual Staffeln within a Gruppe.

As in the previous method, Staffeln aircraft carried numbers 1 - 12 but these were in the individual colours of the Staffel and completely replaced the former Geschwader colours detailed in Chapter 1. The colours used were: 1st Staffeln — 1, 4, 7 — white; 2nd Staffeln — 2, 5, 8 — red; 3rd Staffeln — 3, 6, 9 — yellow.

In cases where a fourth Gruppe was added to a Geschwader the Staffel numbers added were 10, 11 and 12, these being added one to each existing Staffel and were therefore coloured white, red and yellow respectively, but if a fourth Staffeln was added these aircraft carried black numbers.

Fighter codes retained a simple numerical sequence which was officially supposed to be painted forward of the fuselage Balkenkreuz but there were exceptions to this, one notable one being the aircraft of 1/JG1 in which tradition died hard, and the

A Bf 109E-4 which came to grief in Holland when Unteroffizier Hans Schubert had a tyre burst as he was taxying. This aircraft is in a typical Battle of Britain scheme with splinter camouflage on the wings and tailplanes and the hard demarcation line between the top decking colour and the underside blue (Hans Obert via Martin Windrow).

This Me 262A-2A is of 1/KG51. It has a wave mirror camouflage with white tips to the fin/rudder and nose. Of particular note are the bomb racks under the nose, just forward of the wing leading edge (Windrow/Creek).

identifying codes were painted on the sides of the engine cowlings in the position formerly occupied by the old style numbers.

The positioning of the identifying numbers forward of the Balkenkreuz meant the repositioning of the Gruppe symbols and these were moved aft of the cross. The Gruppe symbols were a modified form of those previously applied with the addition of a new symbol for the fourth Gruppe where it existed. The symbols for Geschwader and Gruppen Staff aircraft were also modified at this time and were usually painted in black or white, whereas on Staffeln aircraft they were painted in the Staffel colour. The symbols for individual Gruppen were: I Gruppe — No symbol; II Gruppe — A horizontal line; III Gruppe — A horizontal wavy line; IV Gruppe — A small cross or circle.

The horizontal wavy line of the III Gruppen was to be replaced in 1941 by a vertical bar, but examples of the earlier marking continued to be seen on III Gruppe aircraft long after this, and indeed there are recorded sightings of its use as late as 1945.

Two other symbols that could be seen during the early days of World War 2 were those carried by fighter-bomber elements and ground-attack units. In the former case some Jagdgeschwader had a Staffel equipped for the fighter-bomber role and such aircraft carried a white bomb silhouette aft of the Balkenkreuz in the normal position for the Gruppe identifying symbol. The original ground-attack or *Schlachtgeschwader* units were identified by a black triangle which was painted aft of the fuselage Balkenkreuz, the aircraft being identified by its own number painted in the Staffel colour in the usual position. Once again, it is possible to find exceptions showing the triangle painted ahead of the Balkenkreuz. This identifying symbol was used until the ground-attack units were reorganised in 1943 when they adopted a coloured letter ahead of the Balkenkreuz and normal markings, as in the fighter squadrons, after it.

The system of horizontal chevrons, vertical and horizontal bars, that identified Geschwader and Gruppe Staff officers, were, as already stated, retained and painted forward of the fuselage Balkenkreuz, but to these were added three new designations indicating the Geschwader 1A (operations officer), technical officer and staff Major.

On Geschwader staff aircraft these were the only symbols to appear but on Gruppen staff aircraft the normal Gruppen identifying mark was carried aft of the Balkenkreuz. There were subtle differences between the identifying symbols which enabled the aircraft to be identified as either a

A FW 190A-8 of an unknown unit. This aircraft has a mottle camouflage and the spinner carries a white spiral. The black dot above the III Gruppe marking is a mark on the negative and not part of the aircraft's markings. The style of the III Gruppe symbol is worthy of note as it is not as 'wavy' as shown in official documents thus proving the point that instructions were not always strictly adhered to (Hans Obert).

Gruppe or Geschwader machine. For although Gruppe aircraft carried the Gruppen symbol, in the case of each I Gruppe of every Geschwader where no symbol was carried, it still had to be possible to segregate the aircraft from that of a Geschwader staff officer. There were no equivalent markings within the Gruppen symbols for the operations officer and staff Major.

The structure and organisation of the Luftwaffe was similar, in many ways, to that of the RAF. The basic tactical unit was the Geschwader which was roughly equal to an RAF Group, and within this there were three or four Gruppen which can be approximated to RAF Wings, although in RAF organisation it was more usual to have many more squadrons within a wing. Geschwader staff officers, whose aircraft were marked as described and illustrated, were responsible for the overall administration of the complete unit, whereas Gruppen staff looked after their own elements, instituting directives as laid down by the Geschwader staff as well as those peculiar to their own particular Gruppen.

Each Gruppen was made up of three or sometimes four Staffeln (roughly equal to a squadron), which in turn had a Staffel Kapitan commanding them. These officers carried no special identification on their aircraft but

Defence of the Reich bands

Due to the Allied air superiority over Germany an order was issued to all Jagdgeschwader on February 20 1945, that all aircraft would have coloured bands painted around the small diameter of the rear fuselage to aid identification both in the air and on the ground. The bands were to have a total width of 90 cm consisting of two bands of 45 cm each in the case of two colours and three bands of 30 cm each when three colours were used.

The colours and number of bands allotted to each Jagdgeschwader are shown. There is some doubt as to whether or not this instruction was followed as there are very few examples of aircraft seen carrying these. The noted expert on Luftwaffe camouflage, Herr Karl Ries, claims that in 1944 all fighters engaged in the Defence of the Reich carried a single broad red band, as shown for JG1, and in 1945 this was changed to a blue/white/blue combination, as shown for JG300. It is very likely that Herr Ries's theory is in fact correct. However, since the order was given it is considered worthwhile including full details as technically speaking units should have obeyed this instruction. The general term for aircraft engaged in the defence of the Reich was Home Defence Command (Reichsverteidigund).

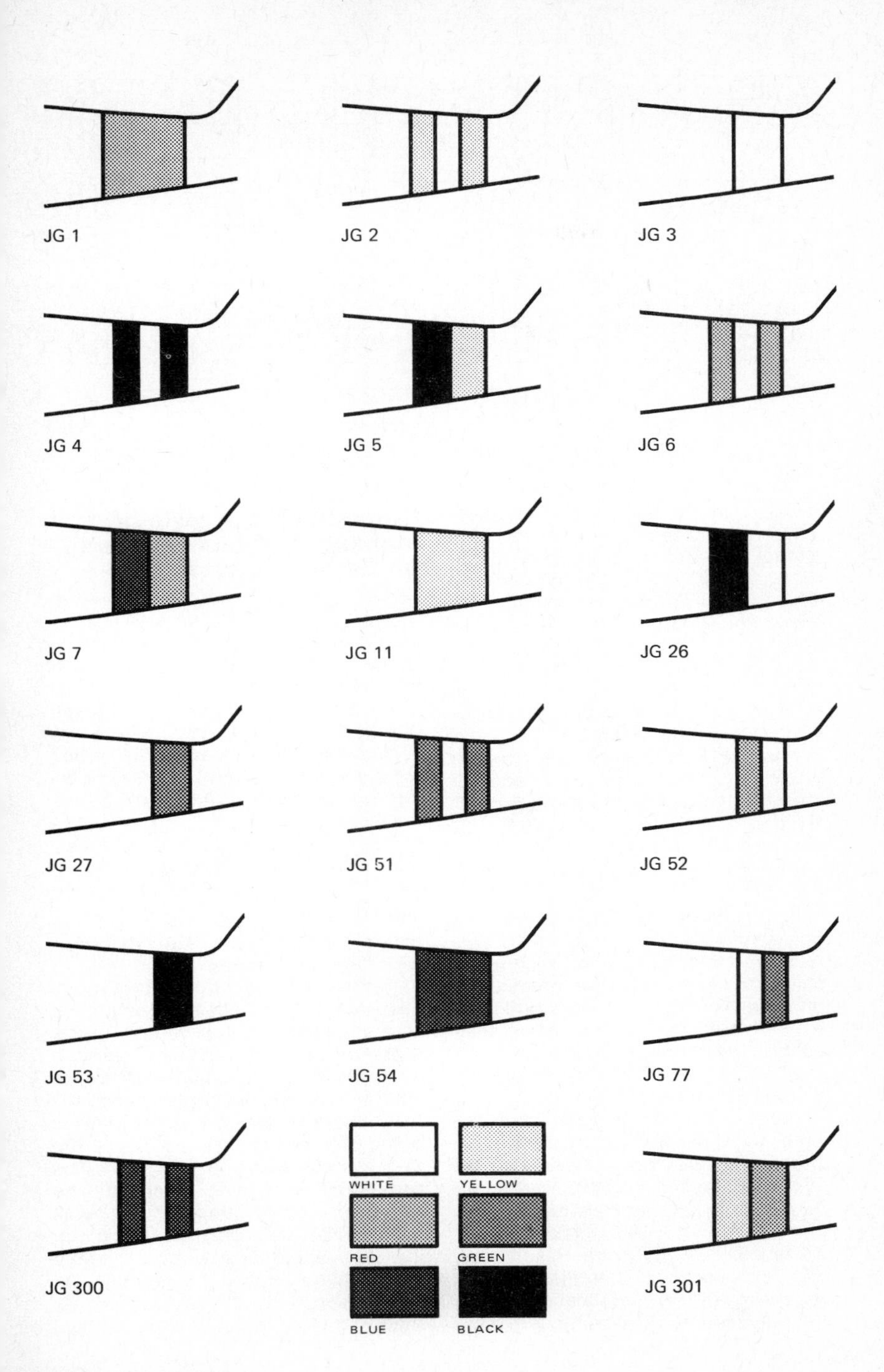
JG 1
JG 2
JG 3
JG 4
JG 5
JG 6
JG 7
JG 11
JG 26
JG 27
JG 51
JG 52
JG 53
JG 54
JG 77
JG 300
WHITE
YELLOW
RED
GREEN
BLUE
BLACK
JG 301

it was usual for them to fly the number 1 aircraft of their Staffel. An average Staffel had between ten and 16 aircraft, giving a strength of between 90 - 150 aircraft to a Geschwader, but these figures should not be taken as officially specified ones as there were many variations. Each Staffel was also broken down into units called *Ketten,* roughly equivalent to a flight, these comprising three aircraft.

Therefore a typical Geschwader could be built up as follows: Kette = 3 aircraft; Staffel = 3 Ketten = 9 aircraft; Gruppe = 9 Ketten = 3 Staffel = 27 aircraft; Geschwader = 27 Ketten = 3 Gruppen = 81 aircraft plus the Staff aircraft.

Aircraft within a Staffel sometimes operated as a *Schwarm* which was four aircraft, or as a *Rotte* which was a pair, therefore a further breakdown gives two Rotte to each Schwarm. This type of breakdown more often than not applied to fighter aircraft although it could of course be applied to any type of Geschwader.

Towards the end of 1938 when the new system was introduced — although it was not generally to be seen until the spring of 1939 — the red band on which the Hakenkreuz was painted on the fin/rudder, disappeared. For a while the Hakenkreuz was retained on a white disc but this gave way to direct application on the fin only, the rudder being left plain except for various 'kill' markings which will be discussed later.

The time had also come for a modification in proportions and positioning of the under and top wing crosses, those on the undersurfaces having a much wider white outline within the centre and black border, and the height being placed equally either side of the wing centre-line, with the centre of the cross being at mid-span. The top surface cross retained the narrower white border but was mounted 200 cm from the wing tip but, like the undersurface one, spaced equally about the wing centre-line. On newly issued aircraft this directive was rigidly adhered to but those re-marked

Proportions and style of alphabet and numbers used on German aircraft

All letters and numerals were drawn in proportion to each other and to their height using a fixed grid as shown.

The height of the letter or numeral was 6/10ths of the adjacent fuselage cross, and was divided into a seven unit grid as drawn. Most letters are 5/7ths wide but wide or narrow letters such as A or T vary, the examples quoted being 6/7ths and 4/7ths respectively.

From this table it is easy to determine the correct style and size of lettering for any type of aircraft, be it large or small.

in the field often showed different interpretations of the size and positions of all national markings, not just the wing crosses.

This type of inconsistency appears many times in the application of camouflage schemes, and where a particular example is worth quoting this will be done, but for the sake of the modeller who wishes to find his own anomalies applicable to the particular model he is making, the majority of finishes and colours mentioned will be those officially laid down by the authorities concerned.

In the early days of the campaign in Poland the size of national markings varied quite considerably and it was not uncommon to see some Bf 109s with underwing crosses occupying the whole wing chord including the ailerons. Similar discrepancies also occurred on the top surfaces and it can only be assumed that such oversize crosses proved a positive aid to identification both to ground and air observers.

The work carried out by the RLM during the late 1930s in the preparation of camouflage schemes was thorough and exhaustive and culminated with the issue of specific patterns and paint colours issued as part of the Technical Manuals that covered the aircraft concerned. The purpose of these instructions was to give an official guide to those concerned when it became necessary to repaint aircraft, thereby maintaining a modicum of standardisation in exterior finishes. In most cases units responsible for maintenance and repair followed the manuals implicitly, but such adherence to officialdom could only be maintained as long as the units concerned had supplies of the correct paints.

So it is not surprising that by mid-1940 non-standard schemes and colours started to appear, especially in cases where work was carried out in the field at unit level. It is logical that combat requirements had to take priority over what, after all, can be considered to be an idealistic set of instructions. Few commanders would have been happy to accept the unserviceability of a particular type simply because the correct colour of paint was not readily available in which to repair its camouflage. Thus the increase of various 'unofficial' colours and schemes, sometimes confined to small areas but

Bf 109Es of 11/JG54 in Russia. The snakeskin style camouflage pattern is dark green and black-green with light blue separations. The rudders and engine cowlings are painted yellow and the nearest aircraft has a badly weathered fuselage cross (Hans Obert).

not uncommon to complete airframes, had to be accepted.

This makes it very difficult for the modeller to be proved conclusively wrong — or for that matter, right — should he choose a shade or colour that is in variance with what might have been stated in a long-forgotten official document. This does not mean that anything can be accepted as being probably correct, for even those involved in unit repairs would have attempted to use a colour that was as close as possible to that specified, and would have been most unlikely to have used a light grey for example, where a dark green was specified or existed. In such cases where a close approximation was just not available a standard primer would be used or, if the area concerned was very small, left in its original state.

Immediately prior to the commencement of hostilities, most Luftwaffe aircraft carried some form of camouflage pattern. Contrary to the belief held by some people that such patterns were left at the total discretion of the painter concerned, quite the reverse is the truth.

Aircraft manufacturers were issued with a set of Technical Drawings by the RLM clearly defining the schemes to be used when finishing aircraft of the type they were supplying. Each drawing indicated the pattern and colours to be used in plan and side views and included alternative schemes designated 'A' or 'B'. This followed a similar line to that adopted by the Ministry of Aircraft Production whereby RAF aircraft with serial numbers ending in odd digits had a mirror image camouflage pattern to those whose serials ended in even digits. The RLM's schemes did not follow this ruling to the letter since the alternative schemes suggested retained the same camouflage pattern but introduced a colour change on each segment.

The segment, or 'splinter type' camouflage as it is more often called, was plotted on a grid over which was placed an outline of the aircraft concerned. The placing of this outline

A particularly good photograph of a Henschel Hs123A-1 of 8/ScWG-1 photographed in Russia in 1942. Fuselage band is yellow which is probably repeated on engine cowling. Overall finish is dark green with light blue undersurfaces. The black *triangle outlined in white indicating a Ground attack unit can be clearly seen forward of the fuselage cross and the code letter 'H' is red. The aircraft carries a name in script just below the windscreen* (US national Archives).

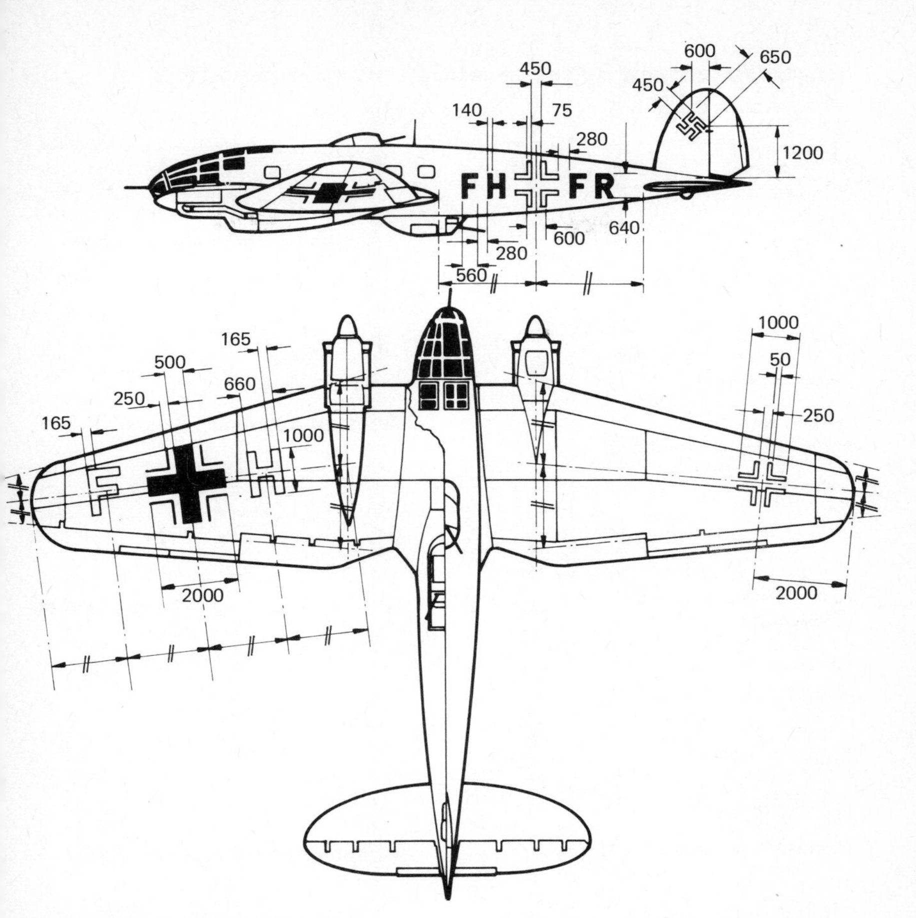

Positions of markings

This drawing, which has been based on official German instructions, shows the laid-down positions, sizes and spacings of national markings and codes. The centre line of the wing is bisected by the wing chord at the nacelle and tip to give the correct datum for the national markings.

The codes are hypothetical and simply show the positions and sizes for the standard four character codes.

All dimensions are in mm.

could result in the splinters not being identical on every type as moving it up or down, to the left or right, produced a slight variation in the resulting pattern. But it was common for aircraft produced in particular batches to have identical splinter camouflage, albeit with some colours reversed. The colours used at this time were Brown (61), Grun (62) and Grey (63) for top surfaces, with undersurfaces in Light Blue (65). It was more common for this camouflage style to be seen on bomber aircraft but examples of both monoplane and biplane fighters carrying it can be found, especially during the period of the operations in the Spanish Civil War.

By 1939 most single-engine fighters of the Bf 109 type were being produced in a simple two-colour camouflage of Black Green (70) on all top surfaces

Organisation and markings of Geschwader post-1938

Geschwader Stab

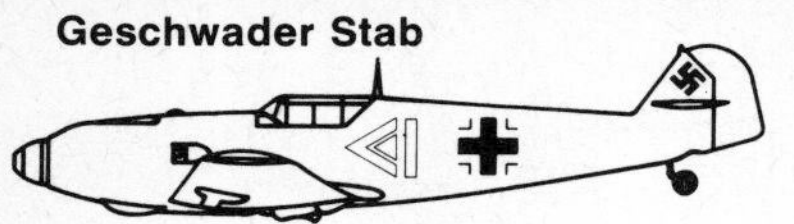

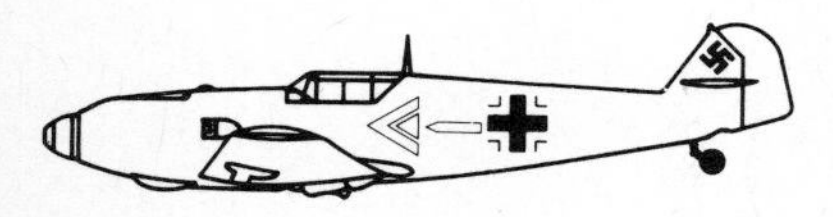

Either of above
Geschwader Kommodore

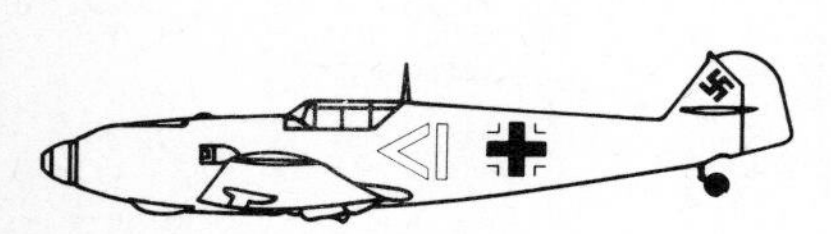

Geschwader Adjudant

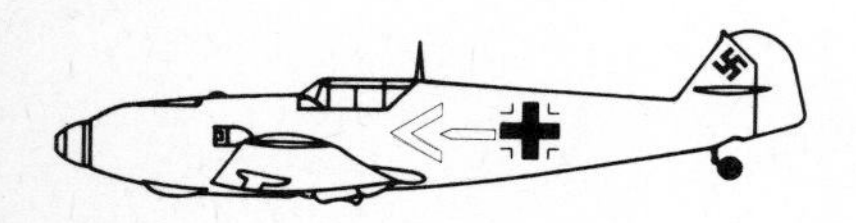

Geschwader Operations Officer

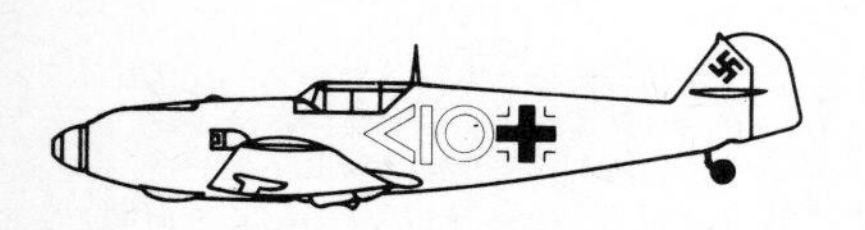

Geschwader Technical Officer

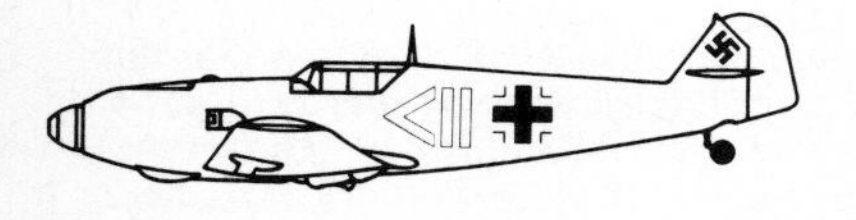

Major Beim Stab

Gruppe Stab

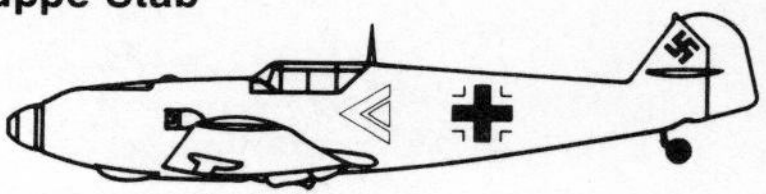

Gruppen Kommanduer I Gruppe

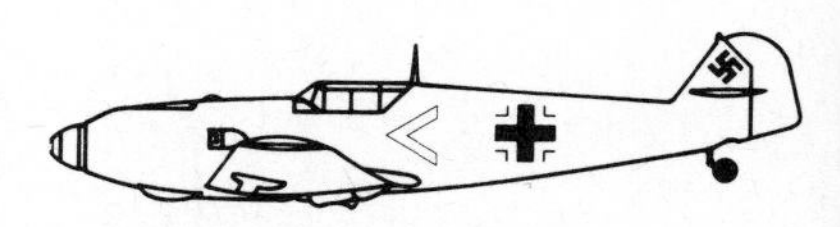

Gruppen Adjudant I Gruppe

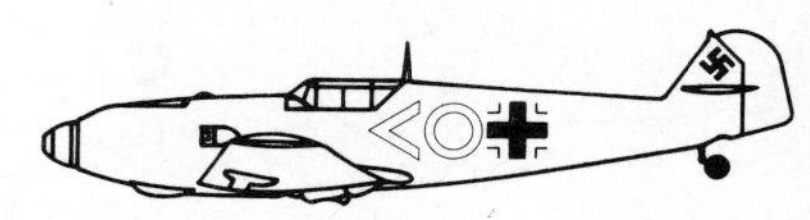

Gruppen Technical Officer I Gruppe

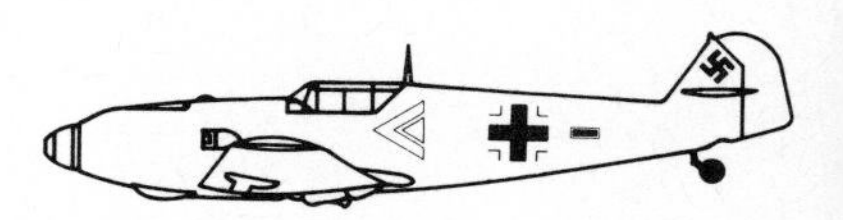

Gruppen Kommanduer II Gruppe

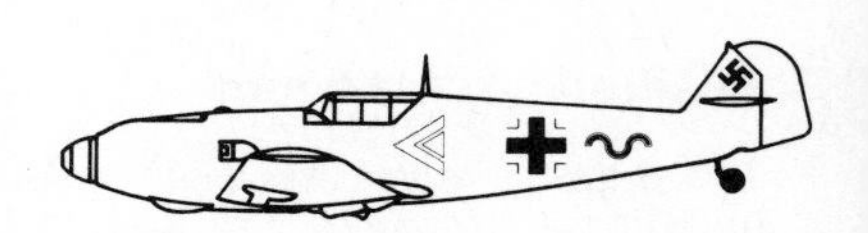

Gruppen Kommanduer III Gruppe

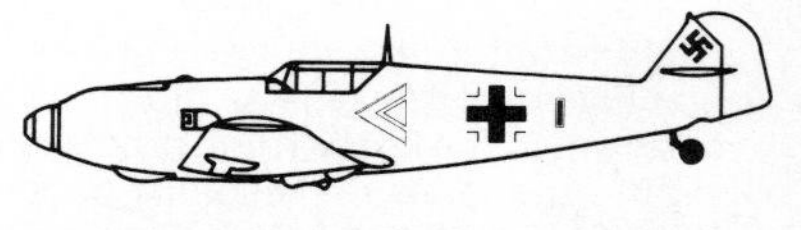

Gruppen Kommanduer III Gruppe (post 1941)

extending right down the fuselage sides, and Light Blue (65) on the undersurfaces. Such aircraft carried the national markings in either the old style or the revised versions already mentioned. Spinners were painted grey or black green but this gradually gave way to the use of Staffel colours either cov-

Staffel

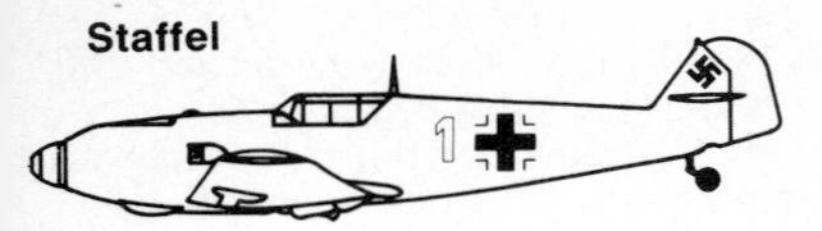

1st Staffel I Gruppe White codes

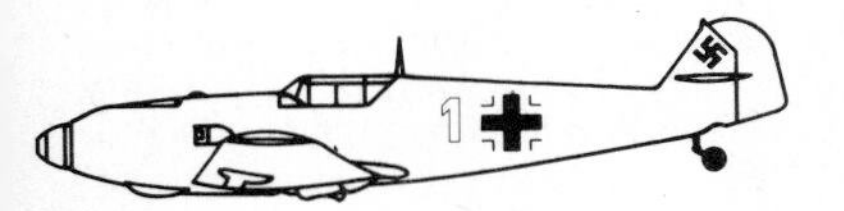

2nd Staffel I Gruppe Red codes

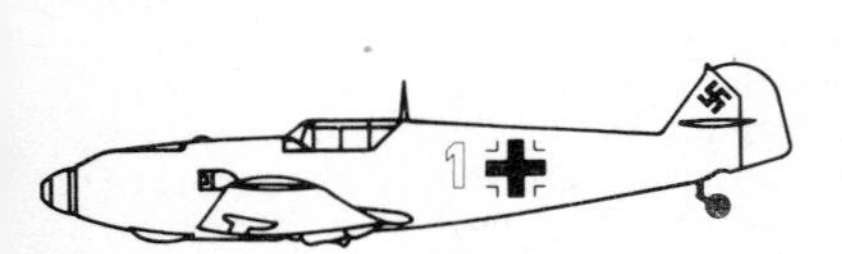

3rd Staffel I Gruppe Yellow codes

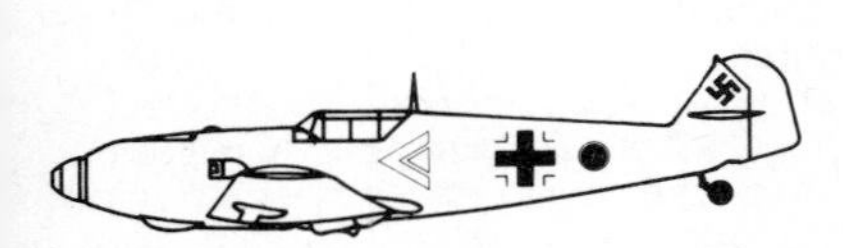

Gruppen Kommanduer IV Gruppe

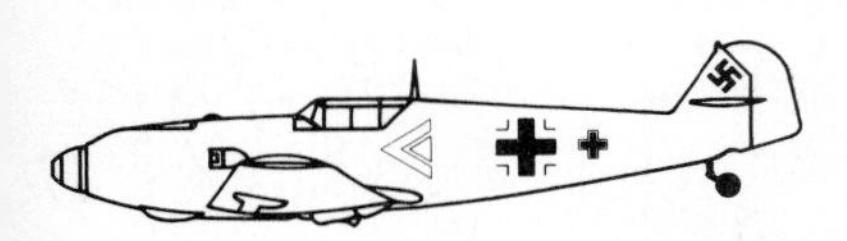

Alternative for
Gruppen Kommanduer IV Gruppe

Jabo (Fighter/Bomber) Staffeln

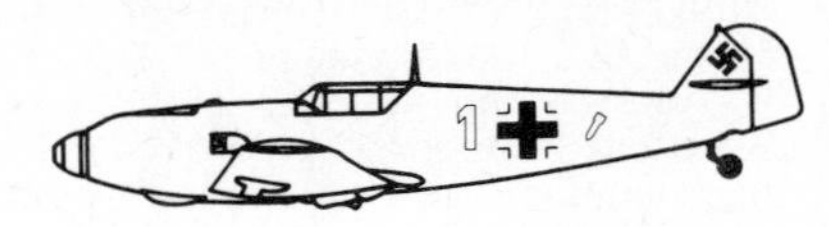

Schlacht (Ground Attack) Gruppen

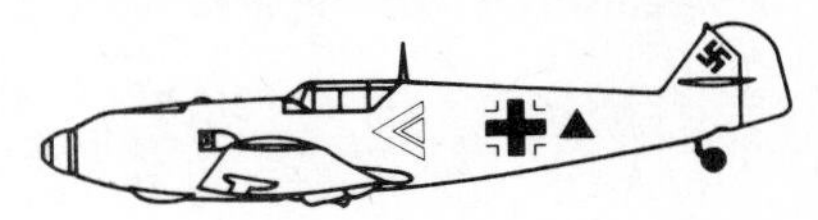

Organisation and markings of Geschwader post-1938

This chart shows the markings carried by Geschwader Staff, Gruppen Staff, Gruppen, and Staffeln aircraft.

The Gruppen Staff of I Gruppe are detailed in full, aircraft of II, III and IV Gruppe would be identified by the Gruppe marking carried behind the fuselage cross as shown in typical examples for the Kommanduer's aircraft of these Gruppen.

Each Staffeln would also carry the marking relevant to the Gruppen to which it belonged, again to the rear of the fuselage cross. So the aircraft illustrated as belonging to the 1st, 2nd and 3rd Staffeln of I Gruppe carry no marking behind the fuselage cross whereas those of the 1st, 2nd and 3rd Staffeln of II Gruppe would carry the horizontal bar, and so on.

Aircraft of a fighter/bomber Staffeln carried the Bomb symbol behind the fuselage cross whilst those in a ground attack or Schlact Gruppen carry the black equilateral triangle. In the latter case it was usual for aircraft of II Gruppe to carry the triangle behind the cross and those of I Gruppe in front of the cross. This often varied.

ering the whole spinner or in a variety of other styles including stripes, half segments, quarter segments or spirals.

The earliest changes to this scheme saw the introduction of a two-tone green on the top surfaces, these being a splinter pattern of Black Green (70) and Light Green (71), again extending down the fuselage sides and retaining Light Blue (65) for the undersurfaces. Throughout the Norwegian and Danish Campaigns the majority of fighters were to be seen in this style of camouflage although some still had the overall Black Green (70) scheme as a legacy from the Polish Campaign.

The third scheme to appear was introduced on those fighters based on the western borders of Germany known as *Sitzkrieger* (sitting fighters), and became common in the early period of 1940 and the first stages of the Battle of Britain. This scheme retained the splinter camouflage of the two greens but the demarcation line on the fuselage became a hard line level with, or just above, the lower line of the cockpit canopy. Below this line the fuselage was painted in Light Blue (65) as was the vertical tail surface and complete undersurfaces of the aircraft.

It was during this period (mid-1940) that the first evidence of unofficial changes to laid-down patterns started to be seen as front line units took into their own hands the repainting of their aircraft. No doubt some of the changes were introduced as the originators of them had their own theories of how a camouflage that was intended to conceal, should look, while others were carried out on the whims of individual commanders or pilots, and still others as readily available supplies dictated.

One of the commonest changes to make its debut was the overpainting of the fuselage sides and vertical tail surfaces in a stipple pattern using both shades of green already employed in the existing splinter camouflage.

The forerunner of the mottle style camouflage that was to be adopted as standard by April 1941, at least on fuselages, was the dark green and grey — probably RLM Grau (02) which was used as a primer and therefore readily available — dense mottle pattern used by some aircraft of 6/JG51. One of the earliest examples of this style was the Bf 109E-4 shot down by a Hurricane of 56 Squadron over the Thames Estuary on August 24 1940.

This early, rather primitive, mottle camouflage was modified in stages that saw the hard fuselage line gradually become a soft line with subtle curves replacing definite angles, a lowering of the demarcation line eventually giving way to this merging into the light blue sides, which in turn slowly changed their colour to light grey.

During the fateful summer months of 1940 changes occurred almost daily in the schemes displayed by Luftwaffe fighters. The standard splinter style camouflage was generally retained on wing and tail surfaces but fuselages became the canvas for many variations and additions. As already stated, the familiar mottle camouflage first appeared in the European theatre and probably derived from re-painting at unit level. The crude stipple style mottle first seen on the aircraft of 6/JG51 gave

This Hs 126 is dark green overall with light blue undersides and belongs to Aufkl GR 41. Full code is 6K+A2 (US National Archives).

Above *A badly 'bent' Bf 110C. Top surfaces appear to be dark green overall and codes have been painted out. It is possible that this aircraft has been used in the night fighter role* (Hans Obert via Martin Windrow). **Below** *Abandoned Bf 110G night fighters of NJG1. Nearest camera is G9+AT which is finished in light blue overall with light grey wave form. The number 160616 on the rudder is the Werke number* (James R. Kingery via R. L. Ward).

way to a more artistic style in which the light blue fuselage sides merged at the base line of the fuselage into light grey which in turn carried a mottle of dark grey or sometimes dark green. Camouflage aside, other markings both of a tactical and heraldic nature began to be seen on many of the aircraft used by the Luftwaffe.

Throughout history men engaged in combat often found a much-needed bond of camaraderie among their fellows, and rivalry, of a serious but friendly nature, soon existed between various units of all types of fighting force. Identity of comrades thus became an important facet and what better way to indicate one's particular allegiance than a badge or emblem? Knights of the middle ages painted their shields with colourful emblems; infantry and cavalry in later conflicts employed banners, flags and badges, so it was natural that when war moved into the air, aircraft belonging to sections, squadrons, wings, and even individual pilots, began to carry some form of badge.

This type of aerial heraldry became essential to morale and cemented a certain esprit de corps among not only those who flew the aircraft concerned, but also those who serviced them. In the 1914-18 war aircraft heraldry adopted mammoth proportions so it is not surprising that during the early days of World War 2 this type of marking, albeit on a much reduced scale, should be revived. The colourful markings of the USAAF and the more subdued forms of identity often seen on RAF aircraft were matched, and indeed, exceeded by the Luftwaffe.

Personal and unit insignia of this type started to be seen in the Condor Legion and many of the badges originated at this time were carried on by the units or pilots concerned during 1939-45. The issue of badges was a complex affair that has no relevance as far as this particular book is concerned. The quantity of them was also such that it would be possible to fill many more pages than are available, but it would be wrong not to mention the markings referred to as they formed a vital part of Luftwaffe lore that is essential to the serious modeller.

Badges were used by Geschwader, Gruppe and Staffel aircraft, so in theory there could be as many as one Geschwader, three Gruppen and nine Staffeln emblems involved. In some cases, units adopted the same basic emblem but changed the background colours to match the Gruppe colour concerned. These badges were most commonly placed on both sides of the fuselage, usually on the engine cowling or just forward of the cockpit, but this was by no means a hard and fast ruling and they could appear in other positions.

The style and form of such badges was to be seen in an infinite number of shapes, sizes and designs, ranging from simple shields to extremely complex geometric patterns executed with great skill and thought. It was not unusual for the Geschwader and Staffel emblems to have motifs that depicted the function of the unit concerned, examples being the bomb sight superimposed over a map of England used by SKG 210, and the hand clutching an RAF aircraft of 1/JG52.

The subtle political humour of some emblems is well worth a study in its own right, and it is not uncommon to find examples wherein symbols associated with British politicians, such as Winston Churchill's cigar, or Neville Chamberlain's umbrella, form a prominent part.

Other used items that could be traced to the location of the unit's origin, or symbols from German folk

Unit badges

There are enough unit badges and emblems to fill a whole book of this size. Nearly every Geschwader had its own badge and in addition to this there were also Staffeln, Gruppen and individual pilot's emblems. Some of them were simple and drab in colour whilst others were nothing short of miniature works of art. A great many depicted items that could be associated with the unit's role, while others took thinly disguised knocks at officialdom and the enemies of the Third Reich.

The examples shown are typical of the shape and content of the majority and serve to illustrate the points made above.

1 *The badge of 1/JG 51 (Mölders). Shield background is silver, diamond is white, goat and hill are black. Surround to shield is thin white line outlined in black.*

2 *The Englandblitz badge carried on night fighters. The diving eagle is white; the flash of lightning carried in its claws is red; the background and countries on the globe are black; the sea on the globe is blue; and the whole surround is white.*

3 *The badge of 1/Schleppgruppe 4. This has a red shield with the caricature of the FW 189 in white with black eyes and mouth.*

4 *The diving crow emblem of 1/StG 1. In this case the body of the crow is black whilst the beak and outline are yellow. The diving crow indicates the dive bombing role of the Stuka and the colours of the beak and surround varied according to the Gruppe, the example shown on the tone drawing of the Stuka on page 50 has a white beak and white surround.*

5 *The shield is the same as badge 1 but this time the diagonal split is a white band on which are three hedgehogs coloured brown with black outlines. Shield background is bright red outlined black. Badge 1V/JG54 (Grün Herz).*

6 *The famous Lion badge of 111/KG 26. The lion and lettering are black and the shield is yellow with a thin black border.*

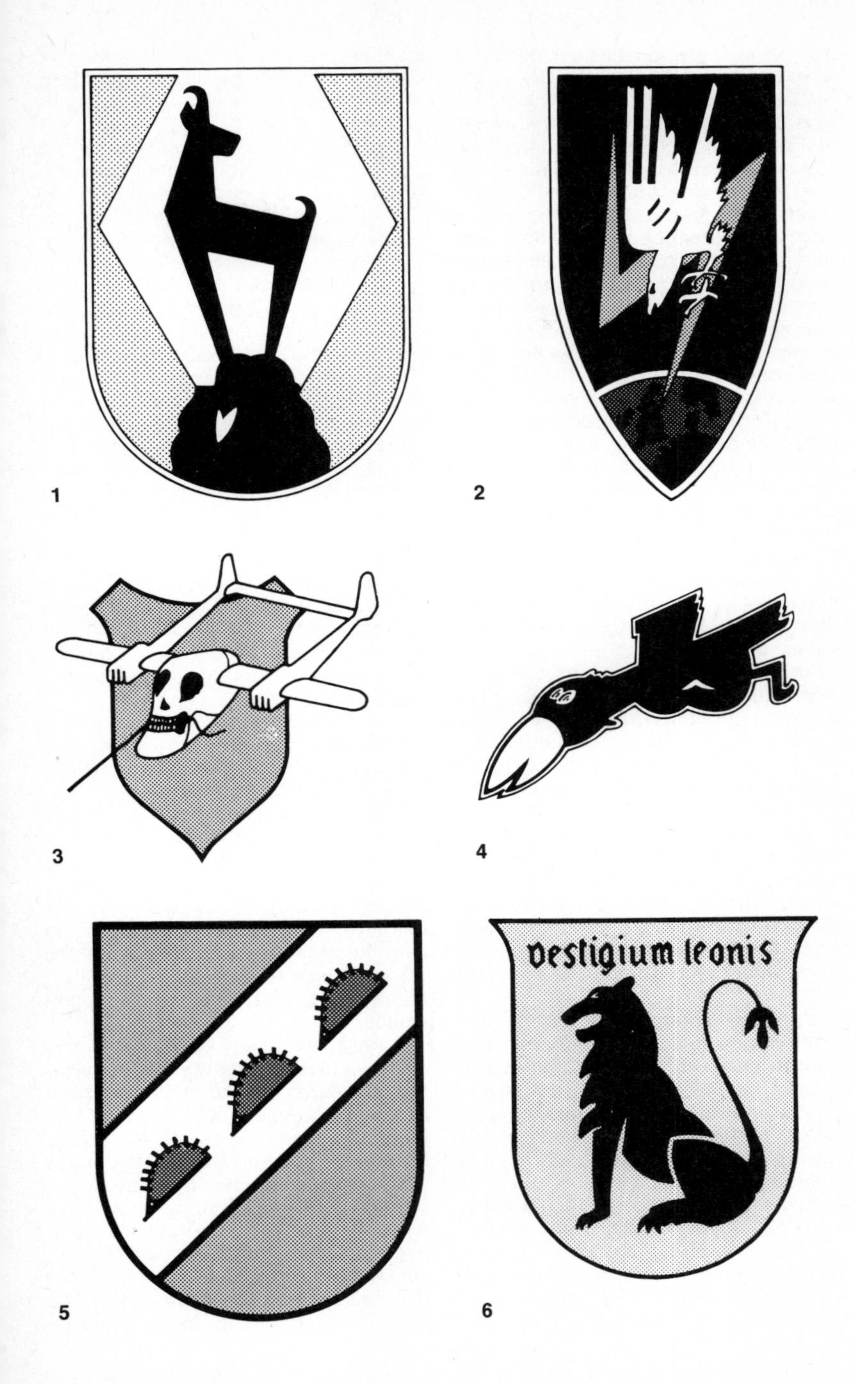

1
2
3
4
5
6

lore. A few representative emblems are illustrated but these barely scratch the surface of the many that can be seen and the reader who wishes to develop an interest in this type of unit marking is recommended to the four volumes of Karl Reis' books *Markings and Camouflage Systems of Luftwaffe Aircraft in World War II* which is a veritable treasure chest of coloured plates illustrating both common and unusual emblems.

The badges mentioned were often supplemented by the personal emblems of individual pilots, which were more often than not painted on the fuselage below the cockpit. These too took many forms, some simply being the name of a girl-friend, wife or sister, while others adopted Disney characters, knights on horseback, family crests, or something with which the man concerned was associated. Such badges were not officially encouraged or discouraged, authority seemingly adopting a non-commital line of thought, preferring to delegate any necessary decisions to unit level.

Individual colouring of aircraft spinners initially followed Staffel colours as already mentioned, but there was a gradual trend, especially among the more flamboyant fighter units, to use this appendage as a further area for colourful attention. The result was a kaleidoscope of patterns varying from simple segmentation in quarters or halves to complex spirals, concentric circles or unequal segmentation. Colours used often stayed within limitations of Geschwader and Staffel, but sometimes more colourful spinners were to be seen.

Speedy recognition of friendly aircraft during specific operations also saw the use of colours as tactical markings. These were mostly confined to engine cowlings and underwing surfaces, and were overpainted on the normal camouflage using water soluble paints. The most common area for single-seat fighters of the Bf 109 series was the motor cowling, the whole of which was painted in a single colour. Similar methods were also employed on the FW 190 and other fighters but it was more often seen on these types, confined to the area underneath the engine, although there are examples of it being used all over or in a variety of patterns.

Painting of areas in colours can be considered as being of a tactical nature, and compared with the invasion style markings employed by the Allies. Any modeller wishing to produce a specific aircraft must be absolutely sure of his facts and research material before he can state quite categorically that the chosen finish, as far as tactical colours are concerned, is 100 per cent accurate. The only thing that can be stated with any model of whatever period, is that the finish used *might* be accurate for any one particular moment in time. During the fervour of operations any aircraft might have areas overpainted in colours that suited the particular duty it was to perform, and such colours could be changed daily or might last for several weeks. So the lesson that must be learned is to get the basic scheme correct but be sure of any additional markings and colours before committing oneself to a dogmatic statement of a general nature.

The most commonly used colour for tactical markings was yellow (04), this was seen on a considerable number of Bf 109s and many contemporary combat reports by Allied aircrew refer to 'yellow nosed' Messerschmitts. These reports mainly refer to the whole engine cowling being painted in this colour as already stated. White was another colour that was frequently used in the same areas as yellow but it was very rare for red to be seen in vast quantities, although it was used to overpaint smaller areas.

During 1941 many single-seat fighters began to appear with white or yellow bands painted around the rear section of the fuselage just in front of the tailplane. The significance of this marking has never been fully explained and it is not proposed to enter into any of the controversial arguments that have surrounded it. General belief is that it depicted the theatre of operations in which the aircraft was involved, white usually being seen on machines operating in the North African, Mediterranean,

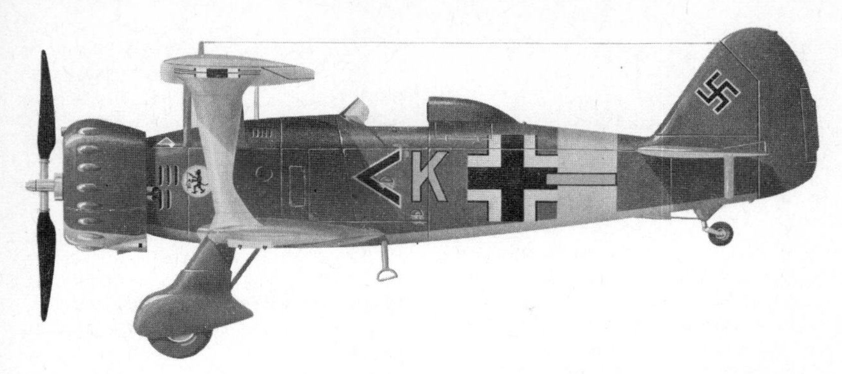

Henschel Hs 123

One of the few biplanes to see active service with the Luftwaffe, the Hs 123 was used in the ground support role. This example was operated by 6/LG2 and has a single colour, dark green, top surfaces with the customary light blue undersurfaces. The chevron is black while the aircraft's code letter and Gruppe marking are yellow. The yellow fuselage band is a tactical marking and the aircraft carries no crosses on the wing undersurfaces.

Balkan or Crimean areas, and yellow on the northern Russian, Scandinavian and Western fronts. These bands should not be confused with the defence of the Reich markings that followed a similar pattern but were introduced, and will be dealt with, later.

The combination of splinter and mottled upper surfaces with pale blue undersurfaces was to remain fairly standard throughout the war, but in 1941 the Luftwaffe was committed to battle on several fronts and geographically orientated camouflage schemes started to appear. During 1941 the Bf 109F started to replace the Bf 109E but it was to be at least another year before the Bf 109E was to be totally withdrawn from all combat areas.

One of the areas where it continued to be used was North Africa where the initial camouflage scheme of light blue undersurfaces (65) was retained but the top surfaces were painted a sandy yellow (79) colour. This tropical finish was applied in the same way as the European finish with a solid demarcation line following the lower line of the cockpit, but it was not long before the fuselage sides were also overpainted in 79, and following the standards set in Europe a mottle of green (80) soon started to appear.

The difference between the tropical and European mottle was that in the former the green blotches extended over the wings and tailplane top surfaces and were not confined to the fuselage sides. These green blotches covered both the light blue fuselage sides and the fin and rudder, so a typical scheme saw the aircraft with sandy yellow top surfaces mottled in green, light blue fuselage sides also mottled in green, and light blue undersurfaces.

The introduction of the Bf 109F to the North African campaign resulted in some aircraft being delivered with the sandy yellow colour covering the fuselage sides but others retained the hard demarcation line three-quarters of the way up the fuselage. It seems most likely that the practice of adding green blotches started when the Luftwaffe went on the defensive in Tunisia and Sicily during 1943, and it was not uncommon to see aircraft that had clearly been transferred from other theatres carrying a hurriedly applied form of tropical camouflage over their former schemes.

Most of the aircraft carrying tropical camouflage also had areas of white on their wing tips, white spinners with the fuselage area immediately aft of the spinner also being painted in this colour, and the white band around the small diameter of the fuselage near the

tail. As in all other theatres they, of course, carried unit badges, Staffel and Gruppe markings as well as individual identification.

The invasion of Russia in 1941 resulted in a large movement of aircraft from the west to the east and most of the early arrivals for the Russian campaign carried the combined splinter mottle camouflage as seen in France and the Low Countries. But it was in Russia that some interesting variations started to appear, among these being the unusual schemes adopted by JG 54 on their 109Es and Fs.

On these aircraft the splinter style was retained on the wings and tailplane top surfaces, but the fuselage either had a solid spine of Schwarzgrün (70) or the two green splinter, while the fuselage sides were covered with a pattern similar to crazy paving. This consisted of irregular areas painted in Dunkelgrün (71) or RLM Grau (02), separated by thin demarcation lines of light blue (65) and streaks of Schwarzgrün (70). Identification areas of yellow on cowlings, rudders and wing tips were not uncommon and the yellow fuselage band was also to be seen. On some occasions the latter, although normally painted around the thin fuselage diameter at the tail, could be seen painted at the mid-fuselage position with the national marking of the Balkenkreuz painted over it.

During the severe Russian winter an overall white scheme for the top surfaces was applied but this was carried out very much at unit level with varying degrees of regularity. The white paint used was a water-soluble type and was applied carefully by some personnel but carelessly by others. The net result was that some aircraft had an even overall finish while others had very much a patchwork quilt effect. On such schemes it was usual for an area surrounding the national insignia to be left in the normal base colour, but how close this was taken to the markings depended a great deal on the dexterity of the man wielding the paintbrush.

As the white paint used was water-soluble it was prone to suffer from the effects of weather very quickly, so it does make an ideal choice of subject for those modellers who like to demonstrate their skill at 'weathering'. In most cases this can be overdone with the result that some models appear weathered to such an extent that a major re-paint is long overdue. But as the white finish was applied over the top of the normal camouflage and was prone to streak and wear off at the slightest provocation, a greater degree of flexibility in finishing can be tolerated. It should be kept in mind, however, that most aircrew and their associate groundcrew, did have pride in their aircraft and would probably carry out a

Messerschmitt Bf 109G

The Bf 109G of the Luftwaffe's leading ace, Major Erich Hartman, who survived the war with 352 victories.

Camouflage is green mottle over light grey fuselage sides with a solid dark green spine to the fuselage. Undersurfaces are light blue and the top surfaces of the wings and tailplanes carry the two-tone green splinter camouflage. The spinner is red with a white spiral and the triangular markings on the cowling are black outlined in white. The heart under the cockpit is red pierced by a white arrow and the name 'Karaya' is in white. The fuselage band is yellow.

new application of the overall white before it began to reflect too badly on their personal attitudes.

Wear along leading edges of wings, tailplanes and fins, plus exhaust streaks, footmarks around entry points, and along edges of removable panels, would be acceptable and more tolerable for longer periods than, for example, large areas of the original camouflage showing through on wings and fuselages. As with all facets of modelling a certain amount of discretion must be applied if a true appearance of authenticity is to result.

Although the splinter and mottle camouflage schemes were to be seen in the majority of cases, some units did adopt patterns that were not unlike the soft wavy lines of the British-style shadow camouflage. In such cases patterns of contrasting dark green and black-green extending over all the upper surfaces were used while undersides retained the familiar light blue.

The introduction of the FW 190 in 1942 saw very few changes in general camouflage patterns, although a relaxation of the strict control seen on early Bf 109s becomes increasingly obvious. The splinter pattern was retained for wings and tailplanes but the demarcation lines between colours was noticeably less hard and did not always follow the normally accepted patterns. The splinter style along the spine of the fuselage more or less vanished for good on this type of fighter, but it was still possible to see some aircraft carrying patterns of two-tone green that could, at a good stretch of the imagination, be claimed to resemble a crude form of splinter. In most cases the underside blue and the top colour merged in a very soft overspray and was mottled over the fuselage sides, producing an almost unnoticeable merging of fuselage colours.

National markings, Gruppe and Staffel markings as well as individual emblems and tactical markings remained much the same as for the Bf 109, but, as stated earlier, it was not usual to see tactical colours applied in such huge proportions as they had been for the Messerschmitt fighters.

The trend in using softer colour demarcations and gentle mottle merging into a mixture of several colours continued throughout the Luftwaffe fighter units until well into 1944, when there appears to have been an almost complete reversal.

During the preceding years two shades of grey, Mittelgrau (75), and Hellgrau (76), had on some aircraft replaced green (62) and light grey (63), although all colour variations continued to be seen until the end of the war, so it was not unusual to have an aircraft painted with the two shades of grey in a splinter pattern and the same colours repeated in a mottle over the fuselage sides.

In 1944 the demarcation line between the fuselage top or spine colour, and the side mottle, appeared as a hard line similar to that seen on the early Bf 109Es of the Battle of Britain period. The line was by no means as set in its position or as straight as it had been, nonetheless it made a return and with it came a much coarser style of fuselage mottle. In a study that is so general in its purpose it would be unwise to make categoric statements as to which aircraft carried which style of camouflage at any particular time, but the variations of this period are brought to the reader's attention to illustrate changes that did occur, albeit not on a vast scale.

With the increase in Allied air cover and penetration after the Normandy landings, and the gradual reduction of air space in which the Luftwaffe could operate, the rapid identification of friendly aircraft became a necessity. With this in mind a system of different coloured tail bands was introduced, although the overall use of these and their allocation is still open to some doubt.

Some writers have suggested that each Geschwader involved in the defence of the Reich was assigned its own colours, while others claim equally emphatically that the colours were assigned to airfields. Whatever the allocation, the fact remains that coloured bands were to be seen and a listing of these is presented in the accompanying drawings. It is possible that not all of these were used, and one historian has gone as far as to claim that it is very likely that only a broad red band, later changing to a combination of blue-white-blue, was ever used. Since it is likely that some fighters did carry this form of identification it has been decided to include it as this guide is concerned only with markings and colours and not the 101 other offshoots that could result from too great a digression into Luftwaffe operations.

In 1944 the RLM were still issuing directives about camouflage patterns and colours, and it is interesting to note that at this time they specified a light grey (76) for undersurfaces on day fighters, with top surfaces being a splinter of dark grey (74) and medium grey (75). Desert camouflage was to consist of undersurfaces in light blue (78), a colour very similar to light blue (65); and top surfaces sand yellow (79) mottled with olive green (80). There is no doubt that some units followed these instructions, but as the air fighting became more of a rearguard action, supply lines became extended, and serviceability dropped, it is unreasonable to suppose that too much care or attention would be paid to 'officialdom', especially in such matters as correct colours or specified camouflage patterns.

The use of a wave-form pattern, as used on some bombers and maritime aircraft, was also used by some single-seat fighter units, but it was more commonly seen on twin-engined heavy fighters which will be dealt with next. But before going on to these and the markings used by Luftwaffe night-fighter units, it is interesting to look at the first aircraft of the jet age that were introduced to combat by German fliers.

A Bf 109E-3 with an overall dark green finish to its top surfaces. The cowling is yellow and the fuselage code '1' is of a particularly large size. The aircraft came to grief on a frozen lake in Norway in 1940 (Hans Obert via Martin Windrow).

The Messerschmitt Me 262 appears to have been covered by no hard-and-fast rules and was to be seen in a variety of schemes. The most common of these was a splinter of two shades of green on the upper surfaces of the wings and tailplanes, with the fuselage spine also repeating the two greens but merging into a mottle on the blue sides, this latter colour extending to cover all the undersurfaces. Many other examples can be found, ranging from a single overall colour to the use of two tones of grey with either a splinter, mottle or wave-form pattern. Staffel and Gruppe markings, when carried, followed the pattern of single-seat fighters.

The Me 163 rocket fighter, which was operated only by JG 400, used a two green (70 - 71) splinter over all top surfaces including the fuselage, and light blue (65) undersurfaces. Once again this was the most often recorded colour scheme but there are examples of overall mottle using green/grey or two shades of grey, and even some with splinter overall except for the vertical tail surface which was mottled in green (71) over its pale blue surfaces. There are also examples of single-colour aircraft but these are rare and the exception rather than the rule.

The short range and limited combat duration of single-engined fighters limited their effectiveness except when they were carrying out their design function of defence over 'home' territory. Since the Germans did not envisage having to carry out such operations in their proposed conduct of the war in which they became involved, it is not surprising that they gave considerable support to the long-range twin-engined fighter. Such aircraft, heavily armed, with

Henschel Hs 129 B-2/R2

Used with considerable success in the ground support role, the Hs 129 was a formidable opponent against armour and had considerable success, especially in Russia.

The aircraft drawn has a two tone green splinter camouflage on all top surfaces with light blue undersurfaces. It is the aircraft flown by Oblt Rudolf-Heinz Ruffer of 8(Pz)SG1. The tactical markings consist of yellow wing tips (undersides only), a yellow fuselage band and yellow nose. The panel immediately in front of the windscreen is dark green and carries the Infantry Assault Badge (Infanterie Sturmabzeichen) which is also painted on the fuselage aft of the trailing edge. The code 'J' and the Gruppe symbol are red outlined in white, and the spinners are, from the front: red and black with a thin white dividing line.

Wing top surfaces have the simplified white outline national marking as does the fuselage, Swastika and underwing markings are standard. The tank emblem 'kill' markings on the rudder are white.

long-range capability and, above all, a long loiter period in the combat area, met their requirements ideally. Fighters of this type could give valuable support to bombers, defending them from the attentions of defending fighters, as well as staying over target areas for long periods to support ground forces.

The concept of such fighters was suited to the type of 'Blitzkrieg' or Lightning war, which was a prominent feature of the operations carried out in Poland and the Low Countries. But they were to find, to their cost, that such aircraft were no match for well flown and highly manoeuvreable single-seat fighters operating over their own territory. A situation thus arose during the Battle of Britain that the single-seat fighter elements of the Luftwaffe, already severely hampered by operating at extreme range and having a very limited air-to-air combat duration, were having to protect the 'heavy' fighters as well as the bombers which should, at least in theory, have been carrying out their assigned tasks under the protective watch of the twin-engined fighters.

The ability to stay airborne for long periods without the worry of having to land for fuel, made heavy fighters an ideal form of defence against intruding bombers, a fact that is borne out by the effectiveness of such aircraft employed in the night-fighter role. But their weakness was underlined when it came to paving the way for bombers through well defended hostile air space.

Reichsmarschall Göring was one of the leading advocates of the *zerstörer* (destroyer) type fighter and was instrumental in pushing the Messerschmitt Bf 110 into Luftwaffe service before it was really ready. The formation of Zerstörergeschwader received such priority that many of the most successful Bf 109 pilots were transferred to the Bf 110 in an endeavour to give the necessary impetus to the units concerned. Despite such high level pressure, very few Bf 110s operated during the Polish campaign, and those that did mainly performed in the ground-attack role.

Increasing quantities were to be seen as the early months of the war progressed but when the aircraft were decimated during the Battle of Britain, their days as long-range day fighters were virtually over. This is not to say that the Bf 110 was not a success, as its true value must be assessed over its ultimate achievements and not isolated campaigns in which it never fulfilled the promise expected from it. As a night fighter, hit-and-run bomber, and an intruder, the Bf 110 excelled and gave the Luftwaffe legion service.

Although classified as fighters, the heavy twin-engined aircraft that came within this category did not carry the same style of markings and codes as their single-seat compatriots, but showed a closer allegiance to bomber aircraft.

National markings were painted in the six normal positions associated with all Luftwaffe aircraft; ie top wing surfaces, bottom wing surfaces, and both sides of the fuselage for the Balkenkreuz, while the Hakenkreuz was painted on the fin, in the case of twin-finned aircraft such as the Bf 110 this was on the outsides of both fins.

A four-character code was carried on the fuselage sides either side of the fuselage cross and for a short period this code was repeated on the under-surfaces of the wings.

The first two characters which were painted before the cross indicated the Geschwader, a full listing of these being given in Appendix 1. The letter-number coding used to define the Geschwader does not appear to follow any rigid procedure, although it has been suggested that there was a rather complicated formula for arriving at these. The codes, which consisted of a number and letter or a letter and number, were usually painted in black, although during the later years of the war other colours were to be seen.

Immediately following the fuselage cross was painted the aircraft's individual letter within its Staffel, and this was usually painted in the Staffel colour or at least outlined in this colour. As with single-seat fighters which followed a numerical sequence commencing at 1, the aircraft carrying

Letter	Designation	Colour of individual a/c letter (third letter in code)
A	Geschwader Staff	Blue
B	I Gruppe Staff	Green
C	II Gruppe Staff	Green
D	III Gruppe Staff	Green
E	IV Gruppe Staff	Green
F	V Gruppe Staff	Green
H	I Gruppe 1 Staffel	White
K	I Gruppe 2 Staffel	Red
L	I Gruppe 3 Staffel	Yellow
M	II Gruppe 4 Staffel	White
N	II Gruppe 5 Staffel	Red
P	II Gruppe 6 Staffel	Yellow
R	III Gruppe 7 Staffel	White
S	III Gruppe 8 Staffel	Red
T	III Gruppe 9 Staffel	Yellow
U	IV Gruppe 10 Staffel	White
V	IV Gruppe 11 Staffel	Red
W	IV Gruppe 12 Staffel	Yellow
X	V Gruppe 13 Staffel	White
Y	V Gruppe 14 Staffel	Red
Z	V Gruppe 15 Staffel	Yellow

The letters G, I, J, O and Q were not used, to prevent any confusion, while the colours assigned to Gruppen were — I Gruppe White, II Gruppe Red, III Gruppe Yellow, IV Gruppe Blue. The V Gruppe, when used, was not assigned any colour.

Focke-Wulf FW 109D-9

This FW 190D-9, which is the aircraft featured in the Airfix kit, is from 9/JG54 Grünherz whose unit badge can be seen below the cockpit. The aircraft has grey (74) and grey (75) splinter on the top surfaces of the wings and tailplanes. The fuselage has splinter camouflage on the top decking gradually merging into black-green mottle over light blue sides. The undersurfaces are all light blue. The spinner is black-green with a white spiral and the code '15' is yellow. This aircraft was destroyed in combat with three RAF Tempests on March 25 1945.

this number usually being flown by the Staffel Commander, the individual letters also usually followed in sequence although the allocation of them appears to have been far more arbitrary than it was for the numerical sequences.

As the chevron and bar symbols used to indicate Geschwader and Gruppe staff were not used on heavy fighters, a system of coloured letters was used for Staff aircraft with the Geschwader commander usually flying aircraft A. The fourth letter indicated the Staffel within the Geschwader or the Staff flight aircraft, this last letter being allocated as follows:

All four characters were painted a uniform size, this being 60 per cent of the height of the fuselage cross which was in fact the standard laid down by the RLM, but variations in size were to be seen in increasing numbers as the war years passed. Positions and sizes of national markings were also quite clearly defined and are detailed on the accompanying drawings.

The four-character codes used by heavy fighter — and bombers — were repeated on the undersurfaces of the mainplanes where they were subdivided into two groups. The Geschwader code was split into two and spaced either side of the cross under the starboard wing, while the individual aircraft letter and Staffel letter were similarly spaced under the port wing. In both cases the characters were applied so that they could be read forward to rear and were 60 per cent the height of the wing crosses. Therefore a code S9+DH would appear S+9 under the starboard wing and D+H under the port wing, it being common practice to use the Staff or Staffel colour in the same way as the fuselage marking. This type of underwing marking was to be seen for the first two years of the war but by mid-1940 a new method started to appear in which only the individual letter appeared painted outboard of the underwing crosses, so in the example quoted this became D+ under the starboard wing and +D under the port. The latter method gradually took precedence over the former although examples of both could still be seen in 1945. Examples of individual aircraft letters being applied to the top surfaces of the wings outboard of the crosses can also be found but most of these appear to be confined to bombers, although some Bf 110s did use the system for a while.

The use of coloured spinners also permeated through to the heavy fighter units but was nearly always confined to Staffel colours and rarely did such aircraft employ the more extroverted styles to be seen on their single-engined counterparts. On some occasions spinners were painted with Gruppe and Staffel colours in stripes so a Bf 110 with a white and red striped spinner could well have been one belonging to the 2nd Staffel of I Gruppe. It is unwise to be too dogmatic about such markings, and generalisation is impossible, therefore unless concrete evidence is available in relation to the particular model being painted, it should not be assumed that simply because the aircraft concerned was operated by the 2nd Staffel of I Gruppe its spinners would automatically be red and white.

Another factor that must be kept in mind when modelling German aircraft is that, on many occasions aircraft were moved from Staffel to Staffel within a Gruppe as combat requirements and losses dictated. It would be quite possible therefore for an aircraft carrying the identifying code of one Staffel to be operated by another for some time before the identity could be changed.

One aspect that is common to all types of aircraft, apart from the national markings, is the use of unit badges which were carried on the fuselage. These could be seen in a variety of positions, the favourite seeming to be below the cockpit or on the nose. Many Bf 110s also carried quite exotic nose art which was sometimes a development of the unit badge such as the large wasp motif carried by the Bf 110s of ZG1, the Wespen Geschwader. The 'sharksmouth' decoration, a favourite with many air forces, also appeared on some Luftwaffe aircraft, examples being the Bf 110s of ZG76.

Variations in general camouflage patterns applied to heavy fighters followed very similar lines to those used by single-seat fighters, but there does seem to be more consistency.

The general tendency was to employ the normal splinter style for upper surfaces with mottled fuselage sides, and although some aircraft were to be seen with the hard line demarcation on the fuselage common to early Bf 109s, these were the exception rather than the rule. The application of the mottle was very much an individual affair but generally speaking this followed what can be considered the normal rule of soft mottling merging into a hard fuselage spine colour which could be either a solid colour or a splinter pattern. This style was common to most fighters used in temperate zones with a gradual replacing of the two shades of green (70/71) by various greys (74/75/76) as the general acceptance of these colours took effect.

Some of the early Bf 110s were painted in dark green (71) over all top surfaces with light blue (65) on the undersurfaces, these being recorded during the early days of the war and during the Battle of Britain. The undersurface blue, either 65 or 78, depending on period, was the most often used colour for heavy day fighters, the changes being confined mostly to top surfaces. An overall top surface mottle of greens or greys became fairly common on Bf 110s which also appeared in the wave-form pattern seen on some single-seat and maritime aircraft. The comments equally apply to other aircraft such as the Me 410 and Ju 88 that were used in the same roles.

As with all styles of camouflage, the main dictum was to be the geographical location in which the aircraft operated, this often resulting in hurried changes to basic schemes. In North Africa the same sandy/yellow (79) was applied to the top surfaces of heavy fighters and this was usually further embellished with blotches of green (80) or dark brown (26/61), although on many occasions heavy fighters operating in tropical zones retained their temperate schemes.

Proportions of National Insignia

The proportions of the two styles of crosses shown here are self-explanatory. In the case of the upper surface cross the small black border is 1:30 of the total width of the cross, the white inset border is 1:20 of the width and the centre black cross arms are 1:4 of the width.

The dimension table for the Swastika is in millimetres and shows the proportions for five sizes of Swastika.

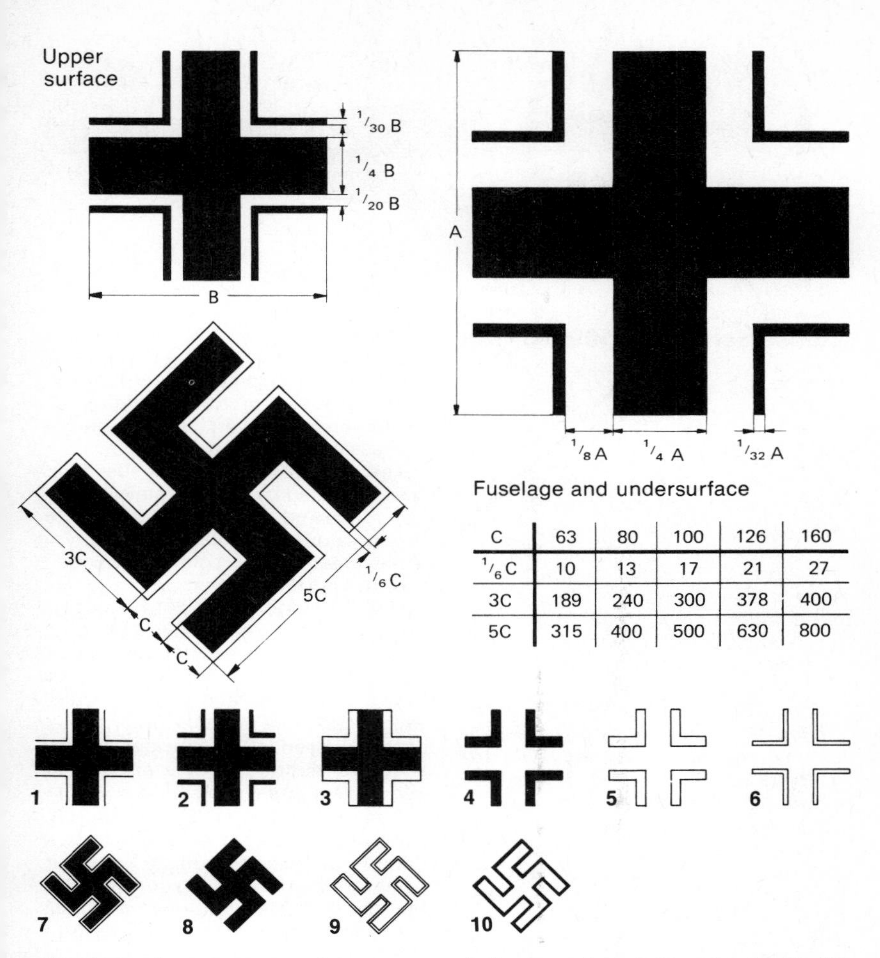

C	63	80	100	126	160
$^1/_6$ C	10	13	17	21	27
3C	189	240	300	378	400
5C	315	400	500	630	800

Styles of crosses and Swastikas used from 1936 to 1945

1 *1936-1943 Upper wing; 1936-1938 Lower wing and fuselage; 1939-1940 Fuselage.* **2** *1939-1940 Lower wing; 1940-1943 Lower wing and fuselage; 1942-1945 Lower wing.* **3** *1942-1945 Lower wing alternative and fuselage.* **4** *1944-1945 Simplified design used on lower wing and fuselage.* **5** *1944-1945 Alternative design to 4 used in same positions, ie lower wing and fuselage.* **6** *1942-1945 Upper wing simplified design.* **7** *1939-1945 Tail fin.* **8** *1944-1945 Tail fin.* **9** *1944-1945 Simplified design used as alternate to 8.* **10** *1944-1945 Simplified single outline used as further alternative to 8 and 9.*

National insignia of the same period were used together but there are many occasions when a mixture of designs can be seen. Individual research into the particular aircraft is the only safe way of being sure of the markings it carried at the time it was recorded or photographed. Markings were usually black/white but grey was used on some night fighters instead of black and on some occasions late in the war, dark green (71) replaced the black on some crosses.

Messerschmitt Bf 109F-4/Trop

This aircraft, which was flown by Hauptmann Hans-Joachim Marseille, is finished in a typical tropical camouflage. All top surfaces are in Sand-Gelb (79) with undersurfaces in Hellblau (65). The demarcation line on the fuselage follows the style of the earlier Bf 109E, whereas some tropical finishes had all the fuselage sides painted Sand-Gelb. The Weiss (21) spinner, forward nose, wing-tips and fuselage bands are tactical markings seen on aircraft usually operating in tropical zones. The yellow 14 indicates that the aircraft is in the third Staffel, and in fact, this particular machine became universally known as 'Yellow 14'. The rudder markings indicate that Marseille had achieved 101 victories, although not all of them were achieved with this aircraft. After his 101st victory Marseille was awarded the swords for his Knight's Cross with Oak Leaves, at which time he was the Staffelkapitän of 3 1/JG27. He went on to record 158 victories before being killed on September 30 1942 when he bailed out of the aircraft drawn but was struck by the tailplane.

The small triangle on the nose is red, the one under the cockpit is a fuel octane triangle and is yellow with the figure 87 inside it, and the trim tabs on the rudder and wings are red. The small cross on the white section of the fuselage cross is red and indicates the location of first-aid equipment. Kill markings on the rudder are all white.

Tactical markings consisting of bands around the small diameter of the rear fuselage and undersurfaces of wing tips were also used on multi-engined fighters and again it is dangerous to be too dogmatic about these. General indication is that aircraft operating in North Africa used white bands and patches, while those in Russia used yellow.

The winter camouflage used in areas dictated by local weather conditions followed an identical pattern to that previously described, whereby a white water-soluble paint was applied over the normal finish.

In 1943 significant changes occurred to markings, the most important of these being to the size of Geschwader and Gruppe codes. As the Allied air power increased and the Luftwaffe was forced more on to the defensive, it became essential to consider concealment of aircraft on their airfields to afford them some protection from the attention of ground attack aircraft. Although the general disruptive pattern of the camouflage afforded some protection it was soon realised that the bold black national markings could be seen without a great deal of difficulty.

The black and white of the Balkenkreuz and Hakenkreuz was replaced by either simple white outlines or black borders, with the normal camouflage as the in-fill colour. Up to this period in time black and white had been the only colours used for these markings but grey and dark green started to appear and, although never universally adopted, were certainly to be seen on some aircraft.

These changes applied equally to all categories of Luftwaffe aircraft but those that were most commonly seen on the heavy fighter units were the reduction in size of the identifying codes of the Gruppen and Geschwadern. This brought about the appearance of very small symbols painted before the fuselage crosses and these were of a size that rendered them completely useless as far as air-to-air identification was concerned, so it is more than likely that they were applied purely for ground handling purposes.

The boast of Reichsmarschall Göering that no enemy aircraft would fly over Reich territory, would appear to have made the provision of any night fighter force an unnecessary extravagance, but in 1940 General Kammhuber laid the foundations of such a force and by the end of the war it had taken a very heavy toll of Allied aircraft. There is little doubt that the Luftwaffe night fighter units were the most efficient both in terms of organisation and ability to be used by any air force of the period. This in no way detracts from the RAF night fighter squadrons who performed a very difficult task, albeit on a much reduced scale, throughout the war years. But the mass raids carried out by the RAF and USAAF forced the Germans into developing the art of night-fighting to a much higher degree than was found necessary by the Allies who never had to cope with massed raids by heavily armed multi-engined aircraft on the same scale as the Germans did.

Before the formation of a night fighter element the task of night defence — if required — fell to a few selected Staffeln from various day fighter Geschwadern, usually flying aircraft that were obsolete as far as the day fighting role was concerned. These aircraft were camouflaged in the standard splinter scheme of the period, their only form of identity as night fighters being a hastily improvised coding of N before the fuselage cross with the individual aircraft identity number following. These codes were usually painted black and outlined in yellow or white. But when the night fighter force was formed in 1940 a camouflage scheme was devised.

Following the pattern adopted by most air forces' night fighters, the Luftwaffe adopted an all-over black scheme for the Bf 110s that initially equipped the night fighter units *(Nachtjagdgeschwader).* The Bf 110 was available in large numbers and had sufficient range, armament and manoeuvrability to make it an ideal counter to the heavy bomber threat. So it is not surprising that when serious thought was given to the formation of specialist night fighter units this aircraft should be adopted.

As the war progressed more sophisticated airborne interception radar, heavier armament and other refinements were added to the Messerschmitt heavy twin, and although versions of the Ju 88, the Heinkel 219 and other aircraft, including single-seat Bf 109Gs, were introduced into the night fighting arena, the Bf 110 bore the brunt of the night defence of the Reich.

Overall black (22) was applied to all surfaces and in many cases was also used to paint out the white area of the national markings, leaving barely distinguishable shapes on the fuselage and wings. The swastika's white outline was frequently left and the unit codes, when carried, were painted grey or red. This style of camouflage was also applied to Ju 88 night fighters during the period although this particular type was outnumbered by the Bf 110.

Contrary to some beliefs the black used was not like the soot black employed on RAF aircraft and was by no means matt, having a slight sheen to its surface that is difficult to reproduce sufficiently toned down on 1:72 scale models. Although black would seem to have been an ideal colour for aircraft whose operations were confined primarily to the hours of darkness, this was not so in practice, and it did in fact show up quite well when subjected to illumination thus negating its prime purpose of concealment.

Later Allied experiments were to prove that a very high gloss black proved much more effective, especially when the aircraft concerned was picked-up by searchlights, but such experiments, as far as it is possible to ascertain, were not carried out by the Luftwaffe.

Apart from the coloured codes the only other form of colour to be seen on German night fighters was the badge of the Night Fighter Arm which was usually carried on the nose under the cockpit and consisted of a white diving eagle superimposed over a red streak of lightning striking at the British Isles. It was also quite common to see a Roman numeral indicating the Gruppe to which the aircraft belonged, painted just behind the badge below the cockpit.

By the end of 1941 new camouflage schemes were introduced on night fighters and these remained in one form or another until the cessation of hostilities. The earliest of these consisted of light and dark grey mottle sometimes applied to the top surfaces of all-black aircraft, but more often than not a return was made to light blue (65) for undersurfaces with two shades of grey on the top.

One of the most widely used schemes consisted of the light blue colouring being applied over all surfaces with the top areas and fuselage sides having a soft dapple style mottle of light grey (76). The demarcation line on the fuselage where the mottle commenced and the density of this mottle was never clearly defined, but, however, it appeared its main purpose was to 'break-up' the aircraft's outline when it was operating over the target area, where illumination from below by ground fires became a serious consideration.

On some aircraft the top surfaces were mottled in light and dark grey with blue retained on the undersides, while there is also some evidence of aircraft being painted light grey overall with dark grey or even black-green mottling. Another variation that was not uncommon was the use of a continuous series of dark-grey lines in a wave pattern over all top surfaces and this scheme is a real test for the modeller if it is to be reproduced authentically.

With the introduction of the new style night fighter camouflage a return to the carrying of full codes was also seen, but towards the closing stages of the war the unit identification was reduced in size as already mentioned. Some night fighters continued to carry the aircraft's individual identity letter under the wings, and on occasions it was also possible to see this letter and the Staffel letter painted under one or both main planes. As stressed before, the variations used over the whole night fighter force were almost limitless, and many examples can be found of aircraft carrying only their code or Staffel letter on the fuselage, no form of identifying code, or a hybrid camouflage scheme overall that bore very little resemblance to those seen in the majority of cases.

One thing that is fairly certain is that tactical colours were not used by any night fighters, although it is possible that day fighters, pushed into the night fighter role on Wild Sau operations, may have carried these as a legacy of their normal function.

Low visibility national markings usually consisting of black outlines or the simplified white crosses and swastikas were the rule of the day for night fighters during the last two years of the war, but again it must be stressed that the conventional black and white markings still lingered on. The use of light and dark grey for national markings was also introduced but this was by no means universally accepted and photographs of aircraft carrying these are rare when compared to those with the more usual black or white styles.

four

Bombers 1939–1945

During the secret build-up years of the Luftwaffe, the bomber was developed under the guise of transport and high speed communications aircraft which were painted in the overall pale grey scheme with black registrations as detailed in Chapter 1. Such aircraft were operated as civil airliners and in this context gave the fledgling Luftwaffe crews valuable training in navigation and long-range operations. Some of them were also used for clandestine reconnaissance duties when they flew over known military areas. In 1935 when the new German Air Force was revealed, it became quickly evident that the real purpose of the pseudo airliners was far removed from that suggested by their peacelike operations.

It was not long after the revelation of the Luftwaffe that bomber aircraft started to carry camouflage. Existing aircraft and those being delivered from the factories now carried a four-colour scheme consisting of light blue (65) on the undersurfaces and a three-tone segmented pattern on their top surfaces. The colours used for this pattern were dark brown (61), green (62) and light grey (63), which were applied in a specifically laid down pattern contained within the manual issued with each aircraft.

There were four basic schemes which were plotted on a grid showing the colour segmentations, the position of the plan view of the aircraft over this grid determining where the pattern would fall. It was customary to project the pattern so that it covered the wing span of the aircraft concerned, and since all the projected lines were of a set layout this could and did, lead to variations of the basic schemes appearing on fuselage and horizontal tail surfaces. The four basic layouts can be divided into two styles in which one was a mirror image of the other, so that scheme 1 was an exact reversed image of 2, and 3 was an exact reversal of 4. In addition to this, transposition of two of the main colours also occurred giving a combination of six basic camouflage patterns and colours. The three-colour segmented pattern was applied to all bomber aircraft until late 1938 when a change to the simpler splinter pattern using two colours, black green (70) and dark green (71) was introduced.

Heinkel He 111P-2

The aircraft drawn belonged to KG55 based at Dreaux, Chartres and Villacoublay during 1940, and was used in night operations against the UK.

The finish is black overall with a disruptive pattern of grey on the top surfaces. The black has been used to overpaint the Swastika and parts of the crosses leaving just the white outlines, this should not be confused with the simplified type of cross introduced later in the war. Code letter 'E' is white outline only, and the lion badge of KG55 is carried below the cockpit.

As with the early fighter markings the national insignia was carried in the normal positions with the Balkenkreuz being painted on the fin/fins in a white disc superimposed over a red band. When the red area was removed it was common practice to overpaint this in one of the basic top colours and many photographs exist showing the clear demarcation of the former red band with no attempt having been made to disguise this by a continuation of the splinter pattern.

The five-character identification system previously described was used on bombers, being laid out either side of the fuselage crosses in exactly the same way.

With the change to the two-tone green splinter style in 1938 came the introduction of the more familiar four-character code as described in the previous chapter for heavy fighters.

By 1939 most bombers were carrying the two-tone green camouflage although some examples of the three-colour segmented scheme and an overall dark green (71), were to be seen, and this pattern remained basically the order for bomber units throughout the rest of the war, changes being confined mainly to those dictated by local conditions.

The organisation of the bomber units *(Kampfgeschwader)* was exactly the same as that already detailed for heavy fighters with the same colours being used to identify Geschwader and Gruppe Staff, as well as Staffeln. So it is safe to say that apart from a few specialist units such as meteorological, reconnaissance and experimental flights, only the single-seat Jagdgeschwader employed the use of chevrons, bars and other symbols as a means of identification.

Bombers used in the first offensives in late 1939 and early 1940 were nearly always finished in the two-tone green/light blue scheme, variations being to colour demarcation lines and slight variations in the splinter layout. However, during this period there were some interesting additions to be seen, as far as identifying symbols and marks were concerned.

Some aircraft started to be recorded with a smaller version of the top surface wing cross painted outboard of the normal one on both main planes. This could be the result of poor aircraft recognition on the part of fighter pilots — a malady that regrettably also affected RAF pilots and resulted in several tragedies where friendly fighters shot down each other or their own bomber colleagues — and one can imagine bomber crews not wish-

Camouflage grid for splinter scheme (Bombers) — facing page and overleaf

These drawings show the variations of the three colour splinter schemes used from 1936-1940 on bomber aircraft. It will be seen that 2 is a mirror image of 1 and 4 is a mirror image of 3.

It should be appreciated that all four schemes were used and that the colours are reversed on identical patterns. ie the colours on 1 and 3 are reversed although the patterns are identical. A further two variations could be achieved by viewing schemes 1 and 3 as mirror images of 2 and 4 and changing the colours accordingly, thus a total of six variations of the basic camouflage using the three quoted colours can be achieved.

It is stressed that the example shown could apply to all types of bomber aircraft and not just the He111 which is illustrated.

The three-colour scheme was changed to a two splinter using dark green and black green (70 and 71) during 1938 and a wide variations of splinters using these two colours were used. They were also planned on a similar grid and are the types usually detailed in kits of World War 2 Luftwaffe aircraft.

The three colour scheme, as shown, was used in the early days of the war and was commonly seen in the Polish and Low Countries campaigns. By the end of 1940 it was very rare although it was still to be found on some second-line aircraft.

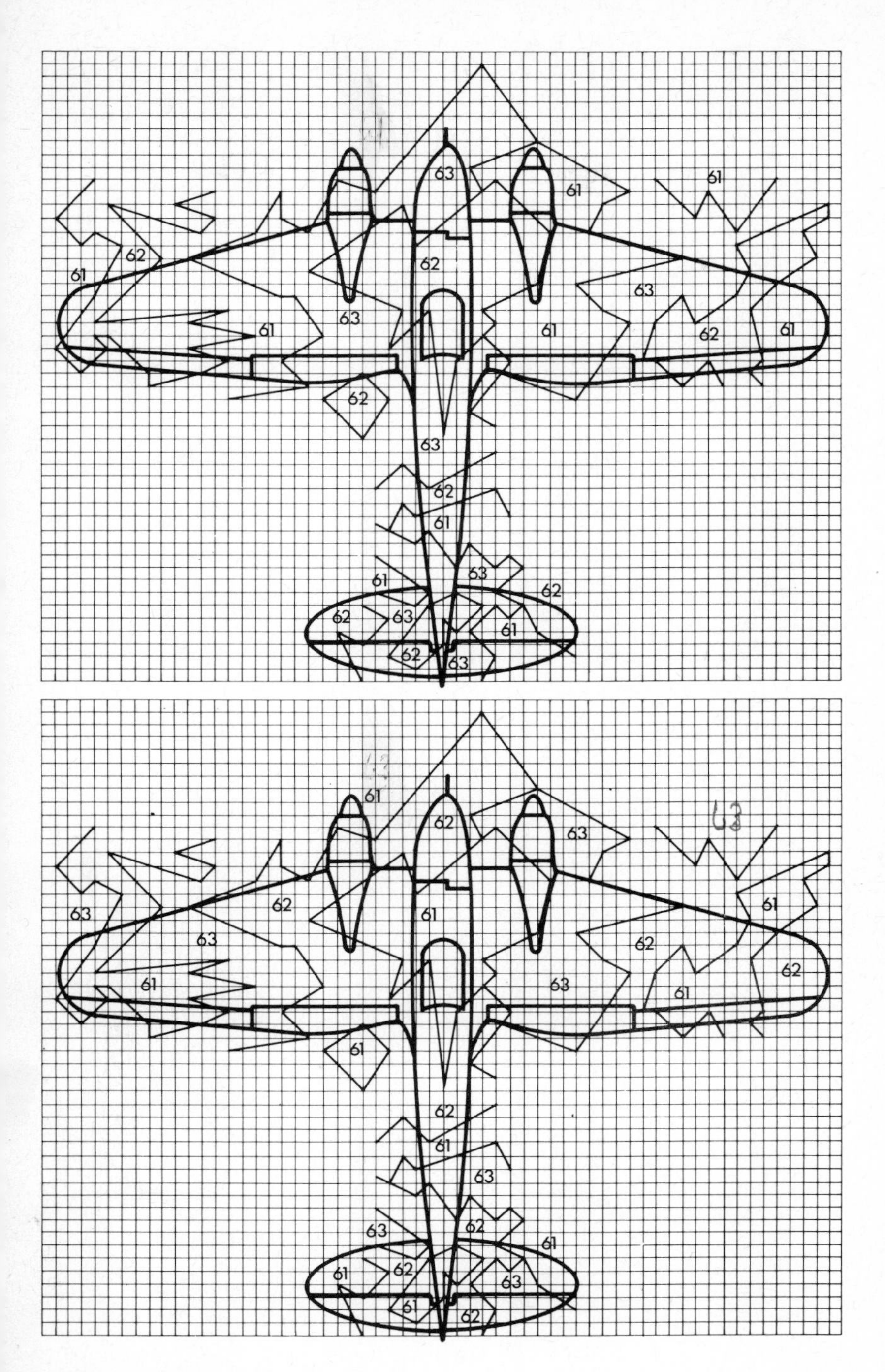
63
61
61
62
61
62
63
63
61
61
62
61
62
63
62
61
61
63
62
62
63
61
62
63
61
62
63
62
61
63
63
62
62
61
63
61
61
62
61
63
63
62
61
61
62
63
61
62

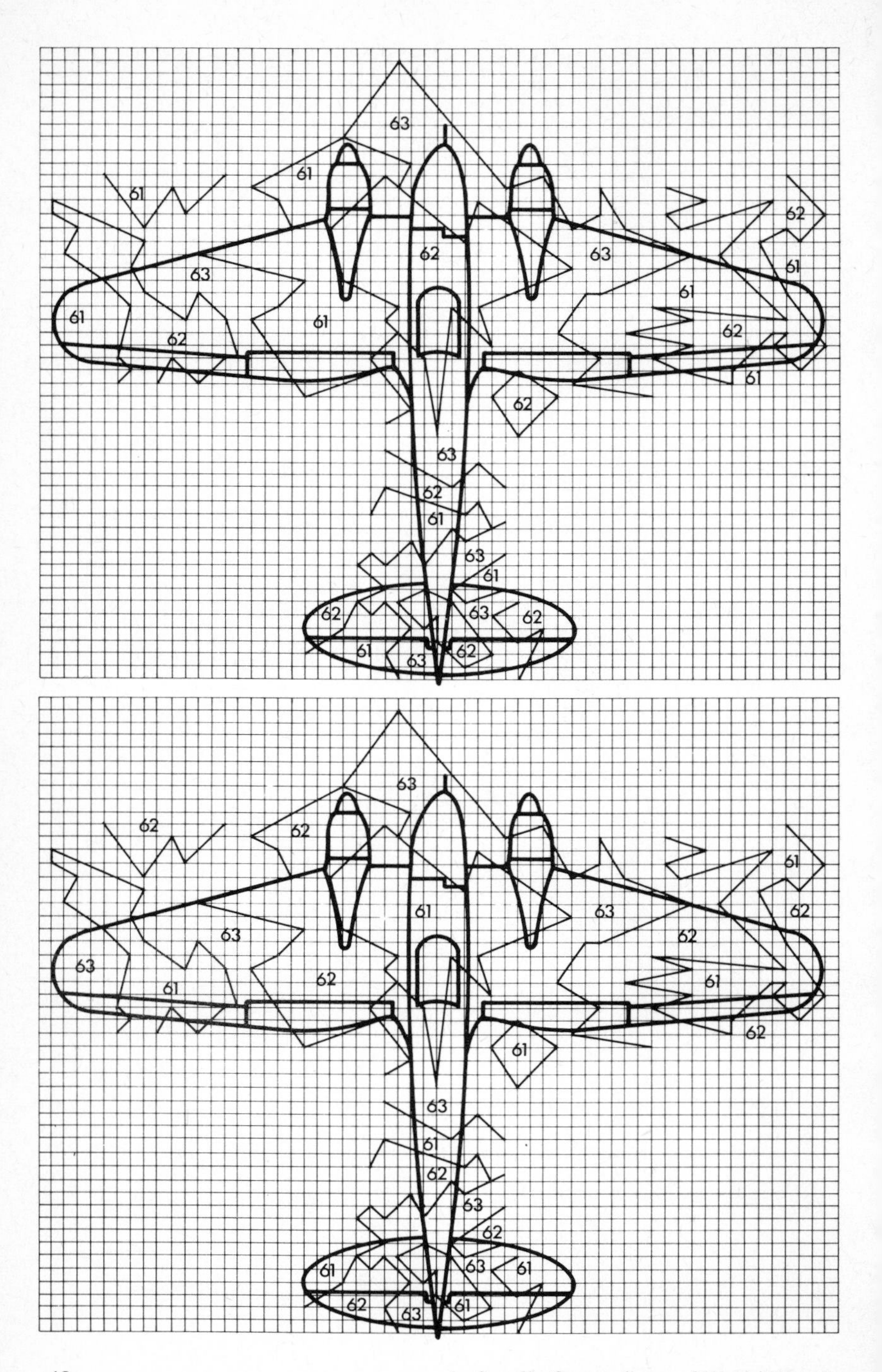
63
61
61
63
62
62
61
63
61
61
62
61
62
61
62
63
62
61
63
61
62
63
62
61
63
62
63
62
62
61
63
63
63
62
61
62
61
62
61
62
61
63
61
62
63
62
61
63
61
62
63
61

ing to take any chances deciding that the carrying of two national emblems doubled the odds against mistaken identity.

Other markings that were also employed consisted of bars of white or yellow painted outboard of the starboard top wing cross and on the vertical tail surfaces; these indicating the Gruppe to which the bomber belonged, the number of bars defining the Gruppe number, eg one bar for I Gruppe, two for II Gruppe and so on. As well as providing additional recognition features these were also used as formation-keeping devices and provided an immediate method of recognition of comrades in arms. Rudders and spinners coloured in Gruppe or Staffel markings were also used on some aircraft, presumably for the same purpose.

During 1940 the use of four-character codes under the wing surfaces also started to disappear and was replaced by the more commonly used individual aircraft letter under both wings or the letter under one wing and Staffel identification under the other.

As with other Luftwaffe elements, the bomber crews were equally as proud of their heritage and carried unit insignia on the nose areas of their aircraft, although this tended to fade out quicker than it did with other units after 1943.

The failure of the Luftwaffe daylight bombing campaign during 1940 brought a switch to night raiding and with it a change in colours. The most noticeable of these was the complete overpainting of the underside blue surfaces with, in most cases, a hastily applied water-soluble black paint. This was also used to paint out white areas in national markings and on some aircraft the splinter style upper surfaces had areas of black applied to them in very irregular patterns. In many cases the black finish was used to overpaint large areas of the fuselage and there are some indications that some aircraft carried an overall black scheme as applied to night fighters.

Bomber operations on the Eastern Front involved few changes in the clement months, but with the onset of winter the expected white appeared to give a degree of concealment in areas affected by snow. The yellow, so favoured by the fighter units in this theatre, also appeared on bombers, usually being confined to undersurfaces of wing tips, with the occasional appearance of it on engine cowlings, rudders and narrow bands painted around the nose. In many cases the addition of white completely obscured the wing and fuselage codes, and as the war years progressed and the schemes were initiated each winter, the use of complete codes became less and less.

The variety of methods in which schemes were applied would suggest that there was no officially laid-down pattern, so the blotched green, mirror wave, snake skin, and other varieties of speckle and flecked colours, that were used on most types of bomber aircraft, were probably devised at unit level and applied with varying degrees of enthusiasm.

Bombers operating in tropical zones also displayed a marked similarity to other aircraft similarly employed with the use of sand/yellow forming a major part of their camouflage. This was applied in the fairly predictable way to top surfaces and fuselage sides, sometimes extending to the lower fuselage line but usually ending about mid-way down the fuselage. This colour was frequently covered in green blotches and, as with the fighter elements, white fuselage bands and areas under the wing tips, as well as the engines, were to be seen on a high proportion of the bombers operating in these zones.

By the end of 1943 the adoption of greys instead of the two greens changed the appearance of bomber aircraft and at the same time the use of the individual identifying letter on the fuselage only, together with the same letter plus the Staffel letter in smaller style on the fin, a method first seen in 1942, became more widespread.

A new assault on British targets,

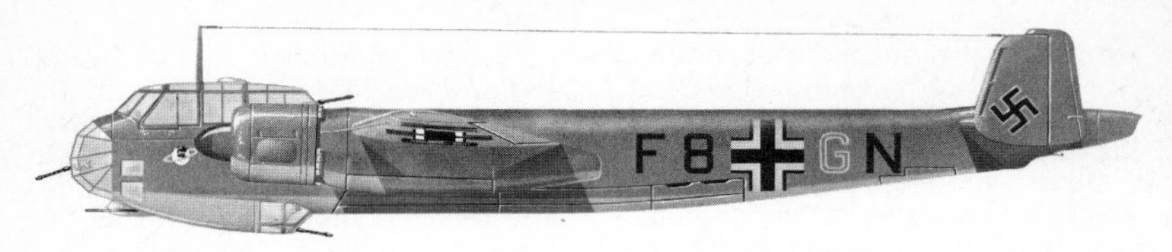

generally believed to be a form of retaliation against the heavy RAF raids of the period, was started in the winter months of 1943/44 under the code name of Operation Steinbock. The weakened bomber units suffered badly from the attention of RAF night fighters whose airbore radar obviously showed no regard for the black and grey camouflage now to be seen on nearly all Luftwaffe bombers. This style of camouflage followed very closely the earlier improvised scheme except that the top surface splinters were now two-tone grey, or a dark grey continuous line wave form over a light grey surface.

Some Ju 88 and Ju 188 bomber aircraft used in this offensive used an overall dark grey with pale blue, or occasionally light grey, disruptive wave form, while others reversed this using light blue overall with dark grey patterns. Aircraft using the former scheme quite often had their codes painted in white outlines only, with the national markings being of the simplified type. A large number (relatively speaking) of Ju 88s and 188s were used as pathfinders or target marking aircraft and most of these employed one of the patterns mentioned. The success of these aircraft can in no way be compared with their RAF opposite numbers, and it would be invidious to draw comparisons, as the tactics employed by the Luftwaffe for night bombing were in no way similar to those of the bomber stream favoured by the RAF.

The use of wave form patterns was a legacy from maritime units and bombers whose operations were mainly confined to strikes against shipping at sea. The effectiveness of them in disrupting the aircraft's outline can said to be proved by the general adoption of them by the night bombing force. The use of mottle was very rare on bomber aircraft and what is generally regarded as a mottle finish was no more than dark grey blotches on a green, blue or black surface. There is very little evidence to suggest that the gradual mottle merging into one or two base colours, as favoured by the fighter aircraft, was to be seen on bombers.

As 1944 progressed the bomber force's fortunes declined and by the end of the year very few units were still in existence as effective fighting forces, and those that were had very reduced strengths. The He 177 continued to attack targets in Russia but there is little significance in its camouflage which was pale blue undersurfaces with the white top surfaces blotched in green.

Many former bomber pilots were hurriedly retrained as fighter pilots in an attempt to stem the increasing Allied air power, and bomber operations were gradually reduced to no more than hit-and-run nuisance raids. But the aircraft used in these raids are of interest because some of them were the fighter-bomber version of the Me 262 operated by KG51. These first jet bombers were usually finished in a standard light blue with a light green and dark green wave line pattern on

Dornier Do 217E

Operating from Bordeaux-Merigna in 1941, this Do 217E belongs to 5/KG40 and carries the codes F8+GN in black with the aircraft's individual letter 'G' red outlined in white. Top surface camouflage is dark green overall and undersurfaces are light blue, this being an example of early bomber scheme variations using a single top surface colour. The badge of KG40 is a globe circled by a yellow band and is carried on the nose of this aircraft below the cockpit.

the top surfaces, carrying simplified national markings and individual identity letters on the fuselage sides.

Another jet bomber that saw limited service was the Arado Ar 234 which was flown by the experienced crews of KG 76. These aircraft had a conventional splinter pattern of light and dark green with pale blue undersurfaces, but the paint used was a high-gloss dope that enabled every last knot of the aircraft's speed to be used to advantage.

Although the Junkers Ju 87 in its wide variety of types was generally considered a dive-bomber and ground attack aircraft it will, together with other aircraft similarly employed, be dealt with in this section.

The Stuka, to give it its incorrect but popular appelation, was blooded in Spain during the civil war where its combat effectiveness was evaluated. During 1939 it brought terror to ground troops and civilians during the Axis advances through Poland and the Low Countries. But although it was decimated by the RAF during the Battle of Britain, it continued to give legion service to the Luftwaffe on almost every front throughout the whole six years of World War 2.

The Ju 87s used in the early days of the war were mostly finished in a two-colour camouflage comprising the usual light blue (65) undersurfaces and black-green (70) top surfaces. Its removal from the Western Front to less hazardous areas around Greece and the Balkans saw only the addition of yellow fuselage bands and engine cowlings, although the latter were also recorded during the Battle of Britain. But it carried as wide a variety of camouflage colours and patterns as its other contemporaries.

In the Western Desert the aircraft was painted in the usual sand/yellow which was disrupted with areas of brown and green, either in blotches or straight line segmentations. In the tropical areas it was not unusual to see aircraft painted in a scheme that closely resembled the RAF shadow camouflage with large areas of wavy edged green or brown painted over the basic sand/yellow. Naturally in Russia the Ju 87 fell into line with most other Luftwaffe camouflage, receiving varying degrees of overall white with flecks and blotches of dark green, yellow fuselage bands — in a variety of positions — and wing tips.

The units equipped with the Ju 87 revelled in a large range of colourful

This He 111H is of KG26 whose badge can be seen below the cockpit. The aircraft has a typical bomber scheme of dark green and black-green splinter on upper surfaces and light blue underneath. The small 'C' on the rudder is of particular interest and is probably the aircraft's individual code letter (Hans Obert via Martin Windrow).

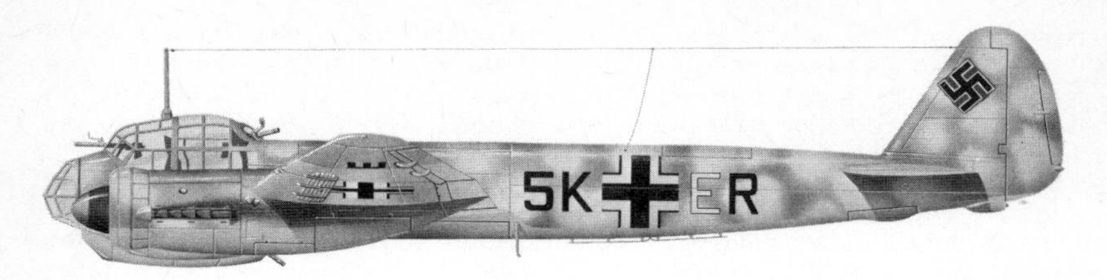

Junkers Ju 88A-4

A Ju 88A-4 of 111/KG3 in Eastern Front winter camouflage. This aircraft has light blue undersides and the original top surfaces are camouflaged in dark green and grey. White water soluble paint has been applied over the top surfaces and has weathered to an extent where the original finish is starting to re-appear. Undersurfaces of the wing tips are yellow and the spinners are white and black.

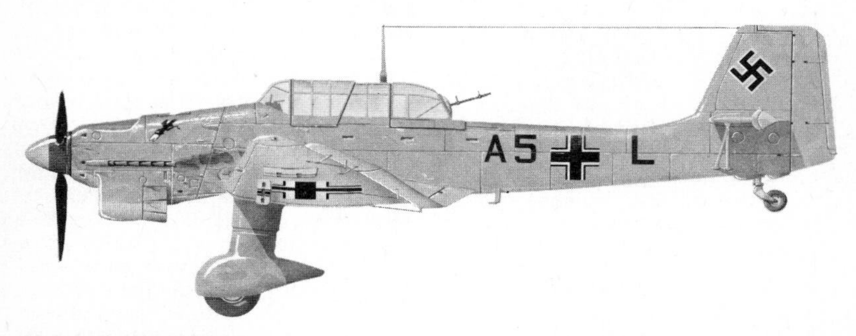

Junkers Ju 87B-2

The infamous Stuka dive bomber is depicted here in a desert scheme as seen on aircraft of 3/SG1. Top surfaces have an overall sand yellow colour with blotches of dark green, and undersurfaces are light blue.

The codes A5+L are all black as is the body of the diving crow emblem of SG1 carried on both sides of the engine cowling, the beak of the crow in this case is white. The scheme shown on this aircraft is typical of the desert style camouflage carried by Ju 87s.

Messerschmitt Bf 110C-4/B

The original dark green top surfaces of this Bf 110C have been overpainted in white soluble paint for winter camouflage on the Russian front in 1941-42. Traces of the dark green show through the white in a form of mottle and there is also considerable wear around access hatches and fuelling points on the top surfaces. Undersurfaces are light blue and the yellow band aft of the wings is a tactical marking.

The codes G9+5N are black except the figure 5 which is red outlined in black. The wasp emblem carried on the nose is from ZG 1 'Wespen Geschwader' who operated this particular aircraft.

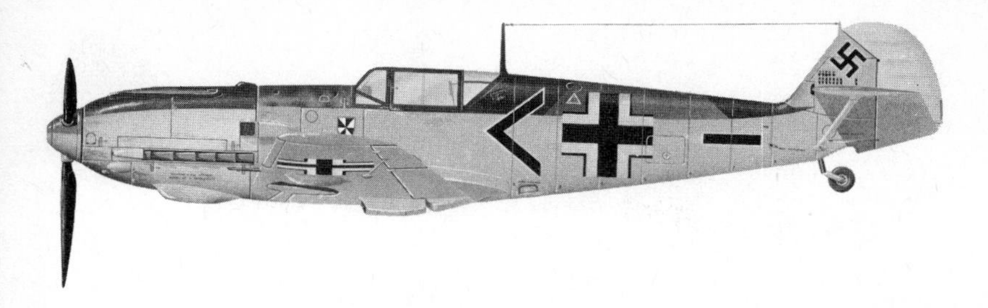

Messerschmitt Bf 109E-4

This Bf 109E is finished in a typical Battle of Britain scheme having black green and dark green splinter on the top fuselage, wings and tailplane with all undersurfaces in light blue. The fuselage sides are also light blue and there is a hard demarcation line between this and the fuselage splinter. The engine top cowling is black green overall which suggests it is a component 'borrowed' from another aircraft. The lower cowling and rudder are painted white in soluble paint.

Spinner is quartered in red/white and 13 kill markings are shown on the fin.

The aircraft is that flown by Oblt Franz von Werra, the Gruppe Adjutant of 11/JG3 'Udet', and it carries the 11/JG3 badge on the fuselage. This is a white shield outlined in red with black segments.

Oblt von Werra was shot down by Flt Lt P. C. Hughes of No 234 Squadron flying a Spitfire on Thursday, September 5 1940, at this time von Werra was flying his 10th mission. The aircaft crash-landed near Marden in Kent and Oblt von Werra was taken prisoner. He eventually became the only German pilot to escape and return to Germany via the USA and was killed later in the war.

insignia carried under the cockpit or on some aircraft on the wheel spats. They also used individual pilot's emblems in similar positions and in this context followed more closely the example of their fighter pilot colleagues. Spinner colours and decorations were also a feature of the *Stukageschwader* featuring the Staffel and Gruppe colours as well as quartered, ringed and spiral patterns. Four-character codes were carried in the same way as the bomber and heavy fighter units with the large characteristic wheel spat being an area that was often use to display the individual identity letter in the Staffel colour.

In addition to the Stukageschwader the Luftwaffe also had *Schlactgeschwader* using fighter-bomber aircraft for ground attack work. On the eastern front two of these units, SchG1 and SchG2, consisting of two Gruppen of four Staffeln each, used a variety of aircraft ranging from the obsolescent Hs 123 to the Bf 109 and Hs 129. The Hs 123 and 129 used a black-green camouflage all over their top surfaces and light blue underneath, while the Bf 109 retained its normal fighter scheme. There were ,of course, variations and the two-tone green splinter was used on the two Henschel aircraft. A distinguishing feature of these aircraft was the equilateral triangle carried on the fuselage showing that they belonged to a Schlachtgeschwader.

The identification of aircraft operated by these units was by a single individual letter in the Staffel colour painted on the fuselage, in this respect they followed the style more commonly adopted by the Jagdgeschwader, although they never carried the four-character coding. The equilateral triangle was painted in black with a white outline although it was sometimes confined solely to the white outline, and appeared behind the cross on aircraft belonging to the IInd Gruppe of the Geschwader with the letter in front of the cross, and on I Gruppe machines the positions were

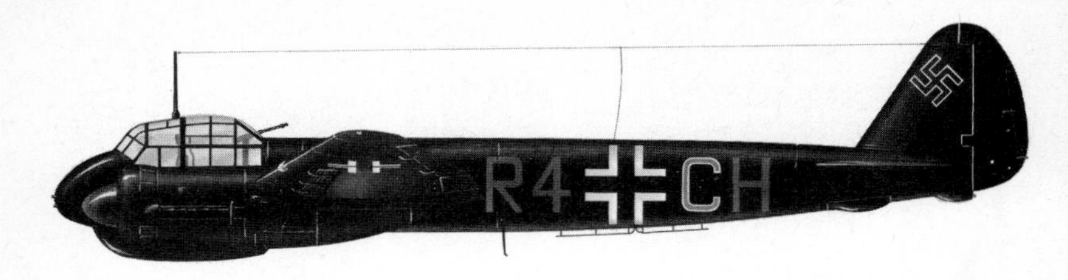

Junkers Ju 88 C-2

The glazed nose of the Ju 88A-4 has been replaced by heavy armament on this C-4 fighter/intruder version. Finish is black overall with national markings in white outlines only. The aircraft belongs to 1/NJG2 based at Gilze-Rijen in 1941 and used on intruder mission over the UK. Codes are light grey with the individual letter 'C' outlined in white. The aircraft carries the Englandblitz badge of the nightfighter arm on its nose below the cockpit. A conversion from the Airfix kit of this aircraft is very simple and was detailed in the May 1974 Airfix Magazine.

reversed. This was a general rule, but as so often pointed out in previous chapters, variations could be seen with aircraft of Staffels from both I and II Gruppen having their code letters forward of the fuselage cross and the triangle to the rear. This is yet another example of the policy laid down by authority being loosely applied at unit level, a situation that naturally increased as the war progressed, and the participants had far more important considerations than the exact positioning of markings on their aircraft.

The Stukageschwader and Schlachtgeschwader were supplemented by two *Schnelles Kampfgeschwader* — literally Fast Bomber Wings — equipped mainly with Bf 110s but having some Bf 109s towards the end of the war, and used to attack ground installations and troop concentrations as well as on hit-and-run missions. The Bf 110s usually retained the normal heavy fighter camouflage with the four-character codes but in later years they could be seen with the single identifying letter, carried by the Bf 109s and other aircraft of the Schlachtgeschwader.

During 1943 the Luftwaffe reorganised their ground attack units, designating them all Schlachtgeschwader which was abbreviated to SG, so the former Stukageschwader, instead of being, for example StG 77, became SG 77, and SchG 1 and 2 became SG 4 and 9, with the SKG 10 (Fast Bomber Wing) receiving the designation SG 10. By the end of 1943 the FW 190 was also being introduced into the ground attack and light bomber role, usually at the expense of the aged and vulnerable Ju 87. Many of these aircraft were hurriedly overpainted with black undersurfaces, or in some case black overall, but they often retained the camouflage of their former units, modified, where necessary, with colours and schemes that suited the environment in which they were operating.

Like fighters and bombers, aircraft employed for ground attack work carried Gruppe and Staffel badges, but in some cases these were replaced or supplemented with the army's combat infantryman's badge. This was an insignia that was awarded to both officers and men in infantry regiments who had taken part in at least three first-wave assaults on three separate occasions and had penetrated the enemies' defences with weapons in hand. The badge consisted of a rifle diagonally across a circle of intertwined leaves topped by the German eagle carrying the swastika. It was painted in white on the fuselage sides either under the cockpit or just to the rear of it, although some Hs 129s also had it applied to the fuselage top decking forward of the windscreen.

five

Other units

As with all air forces, the German Luftwaffe's combat units described in the foregoing chapters, needed support from a host of other aircraft employed in the transport, reconnaissance, maritime and training roles. The diversity of markings used by them was as great as any already described but since it would not be practical, and in any case most modellers seem to prefer to make combat aircraft, they will be dealt with in this chapter in a limited way which will maybe inspire those who decide to add such aircraft to their collections to read more specialised books in which these units are dealt with in greater depth.

These Arado Ar 196A-3s coded 7R+HK and 7R+GK are from 2/SAGr 125 and have a two-tone splinter on top surfaces. Worthy of note is the position of the Swastikas which are on the rudder rather than the fin, a not uncommon position for seaplanes. Fuselage bands are white (US National Archives).

Transport units

The tri-motor Ju 52 was the mainstay of the Luftwaffe as far as transportation of troops, equipment, and the training of paratroopers was concerned, and early in the war units operating these were designated KGzbV which reflected their dual role of transport and bomber operations. The aircraft used were fitted with removable bomb racks and a ventral dustbin gun, the detachment of which increased their fuselage carrying capacity when they were used in their transport roles.

These aircraft supported the advancing German army during the invasions of Denmark, Norway and France when they were usually camouflaged in black-green (70) on their top surfaces and light blue (65) on the undersides. Some aircraft were painted in the bomber style splinter camouflage of two-tone green and all of them carried the four-character code and national markings as applicable to bomber aircraft. Staffel letters were applied in their respective colours and it was not unusual to see Ju 52s with their engine cowlings overpainted in the Staffel colour as well. The central cowling

Seenotflugkommando 1 operated He 59B-2 seaplanes on search and rescue duties during 1940. This aircraft was based at Boulogne and has black codes. The crosses are red and the Swastika is black on a red band. It is of interest that the Swastika is painted vertically rather than the more normally seen displaced position. The Luftwaffe Eagle on the red band is grey, and purports to show the 'unmilitary' nature of the aircraft (via Martin Windrow).

proved a popular place for Staffel emblems which were also applied in the more conventional positions.

Camouflage schemes applicable to the geographical area of operations extended to these machines as it did to most Luftwaffe aircraft, and of course on the Eastern Front the redoubtable white soluble paint was applied just as enthusiastically.

Unit identifying codes were sometimes painted out only to reappear in smaller sizes above the swastika on the fin, and rudders were frequently seen carrying letters or numbers, or a combination of both, that were purely temporary markings used for marshalling purposes on busy airfields.

The original designation gave way to the simpler *Transportgeschwader* (TG) during 1941 but there were no real significant changes to the camouflage colours or patterns. By the end of 1944 many aircraft were pressed into service as transport machines and a variety of codes to be seen varied from manufacturers' codes, to the retention of the unit codes from which the aircraft had been 'borrowed'.

The evacuation of wounded personnel was often carried out by Ju 52 aircraft when their duty as an ambulance aircraft was clearly indicated by the application of a red cross on a white disc on both sides of the fuselage and the top wing surfaces. This replaced the national markings in these positions but on the wing undersurfaces it was painted inboard of the wing crosses. The coloured fuselage and wing tips often associated with a particular area of operations was also applied to transport aircraft, but as with all other markings of this type was by no means obligatory or consistent in size and area of application.

Training units

The marking of training aircraft is a subject that is clouded with some mystery and many aspects of it have yet to be unravelled, the passage of time and the absence of records would appear to make the final solution a remote possibility, but there is always the chance that one day a long forgotten document may appear to add another part to the jig-saw.

Before the rebirth of the Luftwaffe, training was carried out under the thin disguise of civil flying schools whose aircraft carried the normal civil registration commencing with the national identifying letter 'D'. These machines were usually light grey with black codes but were quickly adorned with the red fin band and black swastika on a white disc in 1935. The following year saw the adoption of a five-figure code which was similar in appearance to those carried by front-line units of the same period. They varied in that the first letter was always S denoting *Schule* (school), and the second letter indicated the Luftkreiskommando to which the aircraft belonged, the third letter was the flight, as opposed to the individual aircraft letter, whilst the remaining two numbers were the individual identity of the aircraft within its particular school.

During 1939 training aircraft that still retained civil codes but were to all intents and purposes part of the Luftwaffe had the national letter D replaced by the prefix WL, which was painted forward of the fuselage cross.

Many of the primary trainers used in the war years were painted overall yellow (04) or RLM grey (02) and sometimes carried a white band forward of the tailplane. It goes without saying that national markings were carried in the normal positions and were of the laid-down proportions. One point of interest is that aircraft being flown by pilots carrying out their first solo often carried red streamers attached to the wing tips to indicate to other air traffic that a very inexperienced pilot was in sole control.

Aircraft used for blind flying instruction carried two parallel grey bands around their rear fuselage as an indication of their function.

Many of the aircraft relegated from front-line units to the training schools retained their unit camouflage but this was rarely updated to modern front-line standards and could therefore, at least in the early years, give a fair indication of the period at which the aircraft had been removed from front-line to second-line duties. As the war progressed and the airspace became more hostile, primary trainers lost their yellow or grey finish which was replaced by an overall black-green (70). Training school badges existed and were applied to fuselages and cowlings and it was not unusual to see crests of nearby towns adopted by the schools and painted on their aircraft.

Reconnaissance units

Camouflage colours and patterns used by the reconnaissance, *Aufklärungsstaffeln,* units followed very closely that used by the Kampfgeschwader with all the usual variations applicable to geographical and climatic conditions. The four-figure fuselage codes were common with the exception that the Staffel colour was

A Ju 52/3M used as an ambulance aircraft. Finish is overall dark green with light blue undersides. Fuselage cross is replaced by the international Red Cross symbol which is also repeated on the wing top and bottom surfaces (US National Archives).

rarely used on the aircraft's individual identity letter, the reason being that each Gruppe usually had as many as six Staffeln of reconnaissance aircraft and the defined colours did not cover this number of units.

The Gruppe was the basic tactical unit for reconnaissance aircraft and its number was prefixed by 'F' for those engaged on long-range reconnaissance and 'H' for the short range or army reconnaissance duties. Staffeln operated as completely self-contained units frequently far away from the parent Gruppe, and they could have a variety of aircraft on their strength. Most of the long-range units used Ju 88, Do 17, Bf 110 and Do 215 aircraft, whereas the local units opted for the Storch, Hs 126 and FW 189. There was nothing to distinguish the function of these aircraft from similar types used on other duties, so that a Ju 88 flying on a reconnaissance mission would be identical, in external appearance, to the same type fulfilling a bombing function.

Maritime units

Flying boats operating over coastal waters or at sea were under the command of the Luftwaffe but operated in close co-operation with the Navy, and were perhaps the simplest form of aircraft as far as camouflage colours and markings were concerned.

Above *This Ju 52/3M of 111/KGzbV1 is coded 12+ET and has a two-tone green splinter on upper surfaces with the usual light blue undersurfaces. Once again the unit badge can be clearly seen on the engine cowling* (US National Archives).
Below *A Bf 108 communications aircraft in desert camouflage. Fuselage band and wing tips are white* (US National Archives).

Photographed on Comiso airfield, Italy, this DFS 230B has two-tone green splinter on top surfaces, green mottle on fuselage sides and blue undersurfaces. Code is LH+18 (US National Archives).

Pre-war most of them were painted RLM grey (02) but on the outbreak of hostilities they employed a segmented standard camouflage of two shades of green (72 and 73). By 1941 this had been replaced by an overall one colour green (72) on the top surfaces and this was retained by nearly all such aircraft throughout the rest of the war. Unit codes were of the four-character style painted two either side of the fuselage crosses with Staffel colours sometimes being used for the individual identity letter. Maritime bombers used on anti-shipping strikes used the same style of camouflage but this was overpainted with a continuous wave form of pale blue or light grey from 1943 onwards.

The one colour green top surfaces were also used exclusively by the Arado 196 and He 60 aircraft that were carried on board some battleships for reconnaissance purposes. A considerable amount of controversy arose over the duties of some aircraft that came under the designation of maritime use, but were operated in what was known as *Seenotstaffeln*. These machines were primarily employed as air-sea rescue aircraft whose job was to effect the recovery of downed aircrew.

Aircraft used for this type of work were mainly He 59 floatplanes which were painted white overall and carried civil registrations. The national markings were replaced by red crosses in all positions to denote the unarmed nature of the aircraft but they did carry the swastika on the fin and rudder on a white disc over a red full width band. The civil codes were in black and in addition to being painted either side of the fuselage cross, also appeared on the top and lower wing surfaces. Another symbol that alluded to their non-military use was the Reich Eagle painted on the red fin/rudder band to the top left hand corner of the swastika disc. As a double means of identity and to leave no doubt in anyone's mind of the aircraft's purpose, the red fuselage cross was often repeated on the nose.

These aircraft operated in the Channel and North Sea and claims were made by the Allies that in addition to their normal missions of mercy they were also performing reconnaissance duties and monitoring ship movements. The net result was that the RAF shot down several of these aircraft although their alleged clandestine operations were never proved either way.

In addition to the units dealt with there were other classifications that were usually sub-divisions of existing Geschwadern such as weather reconnaissance flights, courier Staffeln and emergency flights using mainly Fiesler Storch operating over the desert in the same way as the He 59s of the Seenotstaffeln did over the sea. These aircraft were not painted with red crosses but often carried white markings to indicate their function. Generally speaking, aircraft carrying out specific tasks were not differentiated with odd camouflage markings but simply retained the same styles as that employed by the Gruppe to which they belonged.

six

Kill markings

Despite official orders that markings indicating missions or kills were not to be displayed, Luftwaffe pilots, like their opposite numbers, chose to ignore such instructions, feeling that the morale-boosting tallies displayed on their aircraft far outweighed the risk of them being singled out for special attention by enemy aircraft.

Unlike kill markings displayed on Allied aircraft, which were usually confined to the fuselage under the pilot's cockpit, the Germans chose the tail unit as the area to display their successes. As there was no official recognition of this type of marking the styles used varied from pilot to pilot, although a tendency to one particular type eventually began to manifest itself. The area used was the rudder but if the score mounted to such proportions that this proved inadequate the lower portion of the fixed fin also received the artist's attention.

Kill markings in the early days of World War 2 consisted of a white vertical bar without any further embellishments. This practice had started during the Spanish Civil War but kills in that campaign were not carried forward to the 1939-45 World War.

Other methods used in the early days of the war were the painting of miniature national markings of the aircraft destroyed instead of the vertical bars, but there soon developed a consistency in which the white bar was retained but painted above it was the marking of the nationality of the aircraft concerned together with the date of the success.

A method to differentiate between aerial victories as opposed to aircraft destroyed on the ground was introduce by some fighter pilots, and this consisted of the addition of an arrowhead to the victory bar, in the case of an aerial victory this pointed upwards, and for an aircraft destroyed on the ground it pointed downwards. The variation in arrowhead styles ranged from a simple pointing of the white bar to the addition of a broad arrowhead, and colours other than white began to be used. The most significant of these was a black vertical bar which indicated a night victory as opposed to the day victory shown by the white bar. There were variations in this, with some kill markings having a white border or a black diagonal line across the white surface, both indicating night as opposed to day successes. Red was also a colour used by some units as was light grey and no doubt in various operational theatres throughout the combat zones other colours were used to fulfil the same function.

Tail of an FW 200 showing markings indicating attacks on shipping. Just below the Swastika is the legend 'England' followed by ten white bars, this probably indicates strikes in British coastal waters (Hans Obert via Martin Windrow).

The rotation of personnel in the Luftwaffe was vastly different to that used in the Allied air forces and pilots did not complete a number of missions or hours before being rested, they simply continued with their units, taking breaks when leave was due or perhaps doing a spell with training units when they had sufficient experience. In view of this, fighter pilots, in particular, spent much longer periods in combat zones so accordingly had greater time in which to accumulate successes.

The result of this was that some pilots accumulated a large number of victories, all of which could not be displayed in the area available, so a system whereby a large block was used to indicate a block of ten victories was instituted. Parallel to this, decorations awarded to the pilot concerned were also painted on the rudder and included the number of aircraft destroyed when the award was made. This method was only used when the Knight's Cross or a higher decoration was given and under this, or alongside it, the normal kill markings were applied as further victories were achieved.

If an aircraft was damaged and a new one assigned the original victory markings were sometimes transferred but this depended a great deal on the whim of the man concern. Some didn't bother to transfer their kills while others insisted that their scoreboard was displayed on every aircraft they flew.

Naturally every man was proud of the success he achieved and many of them liked to give an indication of the type destroyed, using silhouettes of single, twin or four-engined aircraft, as

Kill markings

The following are some examples of 'kill' markings used in the Luftwaffe.

1 *A general marking used initially in the Spanish Civil War and carried through into World War 2 signifying a victory without any explanation. Usually painted white for a day victory or black with a white outline for a night victory.*

2 *The upward pointing arrow signifies an aerial victory as opposed to an aircraft destroyed on the ground in which case the arrow points downwards.*

3 *The marking shows the nationality of the aircraft destroyed and the date the victory was achieved, in this case an RAF aircraft on June 21 1940.*

4 *This was a marking used mainly by night fighter crews and shows the nationality and type of aircraft as well as the date. In the case illustrated this is a victory over a four engined RAF bomber on July 14, year not shown.*

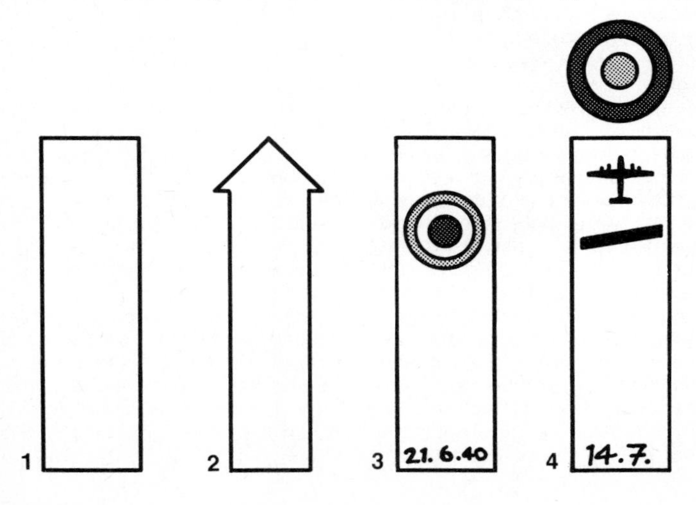

The rudder of this Bf 110G shows ten victories, nine against British aircraft and one Russian. The wiggly line camouflage is dark grey on light blue and the fuselage carries the simplified white cross (James R. Kingery via R. L. Ward).

applicable, painted on the white bar. Others used a system of diagonal bars to show the number of engines of their erstwhile adversary while another, not uncommon method, was to split the white kill marking vertically to give the same information.

Night fighter pilots used the same method of showing their victories but they were particularly proud of successes over the RAF's Mosquito intruder or bomber, as to catch such an adversary, that was faster and more manoeuvrable than their own aircraft, showed a marked degree of skill and warranted special attention in the form of a marking that clearly indicated the type destroyed. The method used was usually to paint 'Mosquito' on the kill bar or on a black diagonal painted across it.

Such markings were not solely the prerogative of fighter pilots. Crews engaged in anti-shipping strikes marked their successes with a silhouette of the ship sunk with its tonnage and the date of the attack painted on it. Ground-attack pilots also used outlines of tanks, railway engines and other appropriate symbols, again with dates and sometimes locations marked on them. Bomber crews indicated missions with vertical stripes but were also more flamboyant on occasions with markings showing factories, or names of targets. Such variety of markings was not solely confined to one particular type of aircraft, for a Bf 109 pilot who attacked a ship might well paint an appropriate silhouette on which was shown the area he had hit.

Ship markings tended to vary a great deal, in the case of a total sinking the ship might be shown going down or as a completely filled outline, while those damaged might be an outline with only the damaged portion filled in. Barrage balloons were also popular symbols, their outlines being used to indicate when they had been destroyed, once again the date of the success was often applied as well.

In addition to showing the operations flown, bomber crews also employed kill markings to indicate successes over enemy fighters, these were the same style as those favoured by fighter pilots and were often painted on the aircraft's rudder or fin but were also to be seen below the cockpit canopy. Various awards for the number of completed operations were also made to bomber crews and again these were often displayed on the aircraft in either of the two positions favoured for mission and/or kill markings.

Not to be outdone by their colleagues, pilots and crews of transport aircraft engaged in the dropping of paratroops or supplies, also marked their achievements choosing, as one would expect, a silhouette of a parachute.

appendix one

Geschwader codes

The following list indicates the codes used to identify various Geschwader and were applied before the fuselage cross as detailed in Chapter 3.

A1 KG53 'Legion Condor'
A2 ZG52 (Ist Gruppe)
A3 KG200
A5 StG1, modified to SG1
A6 Aufkl Gr 120
B3 KG54
C6 KGzbV 600
C8 TG5
C9 NJG5
D1 SA Gr 126
D5 NJG3
F1 KG76 and 1/STG76
F2 Erg(F) Gruppe
F6 Aufkl Gr 122
F8 KG40
G1 KG55
G2 Aufkl Gr 124
G6 KGzbV 2 and TG4
G9 NJG1
H1 Aufkl Gr 12
H4 Luftlande-Geschwader 1
H7 StG3
H8 Aufkl Gr 33
J4 Transpt St 5
J9 StG5
K6 Kü Fl Gr 406
K7 Aufkl Gr Nacht
L1 LG1
L2 LG2
L5 KGrzbv 5
M2 Kü Fl Gr 106
M7 Kü Fl Gr 806
M8 ZG76
P2 Aufkl Gr 21
P5 Sonderstaffel
R4 NJG2
S2 StG 77, modified to SG77
S3 TGr 30
S4 Kü Fl Gr 506
S7 StG 3
S9 Erpr Gr 210, then SKG210 and later ZG1
T1 Aufkl Gr 10
T3 Bordfl Gr 196
T6 StG 2 'Immelmann', later SG2
T9 Luftwaffe trial unit
U5 KG2
U8 ZG26 'Horst Wessel' (until June 1941)
V4 KG1 'Hindenberg
V7 Aufkl Gr 32
W1 Reserved for Me 321 series
W2 Reserved for Me 321 series
W3 Reserved for Me 321 series
W4 Reserved for Me 321 series
W6 Reserved for Me 321 series
W8 Reserved for Me 321 series
W7 NJG100
X4 Luftransportstaffel See 222
Z6 KG66 (Ist Gruppe)
1B 13(Z)/JG5
1G KG27 'Boelcke'
1H KG26
1T KG28
1Z KGzbV 1, later TG1
2F KG54 (until March 1940)
2J ZG1
2N ZG76
2S ZG2
2Z NJG6 (from August 1943)
3C NJG4 (January 1943 - July 1943 NJG5)
3K Minensuchgruppe der Luftwaffe
3U ZG26 'Horst Wessel' (from July 1941)
3W NS Gr 11
3Z KG153, later KG77
4D KG30
4E Aufkl Gr 13
4N Aufkl Gr 22
4U Aufkl Gr 123
4V KGzbV 172, later TG3
5D Aufkl Gr 31
5F Aufkl Gr 14
5J KG4 'General Wever'
5K KG3
5T KSG1 (from February 1943 KG101)
5Z Wekusta 26
6G StG 51, also StG 1
6I Kü Fl Gr 706
6K Aufkl Gr 41
6M Küstenstaffel

6N KG100 (formerly KGr 100)
6U ZG1
6W SA Gr 128 + BORD Fl GR 196
7A Aufkl Gr 121
7J NJG102
7R SA Gr 125
7T Kü Fl Gr 606
7V K Gr zbV 700
8T K Gr zbV 800, also TG2
8V NJG200
9K KG51
9P K Gr zbV 9
9V FA Gr 5
9W NJG101

appendix two

Luftwaffe colours

Up until the formation of the Luftwaffe in March 1935 there were no laid-down camouflage colours, but after this the RLM issued specific details in manuals of colours that were to be used. These manuals were continuously up-dated with the addition and removal of various colours, but to detail all these in a book of this nature would achieve very little.

Generally speaking, most manufacturers delivered aircraft to the Luftwaffe in a overall RLM-grau finish which was in fact a greenish-grey colour, with walkways marked in a 4 cm wide red line. Naturally there were exceptions and in some of these the aircraft concerned were delivered in the official camouflage colours specified.

Since this book is primarily concerned with the World War 2 period, only those colours used during this period are listed in the following table. The original scheme of 61, 62, and 63, dark brown, green, and light grey, was being phased out during 1938 and very few aircraft were to be seen carrying this after September 1939, the colours having been replaced by 70 - 71.

The extensive range of Airfix enamel paints can be used to produce colours that match, or at least are very close approximations to those used on Luftwaffe aircraft. It should be stressed, however, that as in any paint mix, the basic colours of batches can vary — this applies equally to the Airfix paints. It is possible therefore that two tins of for example, M5, bought at different times, may not produce identical finishes.

Paints should be thoroughly stirred before use and the mixes shown measured as carefully as possible, by doing this you will get colours that will produce reasonably authentic finishes. Use the paints as thinly as you can and remember, it is better to use several thin coats than one thick one.

It is worth mentioning again, that many modellers tend to fall into the trap of 'generalisation' far too easily, it is not really possible for anyone to state quite categorically that any model is incorrectly coloured, unless he has in his possession *exact* colour mixes as used in the Luftwaffe over 30 years ago, a situation that is clearly an idealistic dream.

Colour charts of Luftwaffe paints are reproduced in Vol 1 of *Markings & Camouflage Systems of Luftwaffe Aircraft,* also the *Me 109 Gallery,* and in addition to these, the Kookabura book *Luftwaffe Camouflage and Markings,* gives a comprehensive breakdown using modern paint guides. All of these are helpful when mixing colours, but in the case of colour charts it should also be kept in mind that the problems of reproducing accurate colours still presents printers with problems.

Naturally, every effort should be made to paint models as authentically as possible, but there is still a lot to be said for the old adage 'If it looks right, it is right!'.

RLM colours

Number	Type	Main uses	Airfix paint equivalents
00	Wasserhell (Colourless glaze)	Protective varnish	MV1
01	Silber (Silver)	Overall finish on some aircraft. Undercarriage components, interior panels	G8

RLM colours

Number	Type	Main uses	Airfix paint equivalents
02	RLM-grau (RLM-grey)	Used as exterior finish on early Luftwaffe aircraft. Inside wheel wells, cockpit interiors, as primer and overspray on some mottles	2 parts M26 1 part M10
04	Gelb (Yellow)	Identification and tactical markings	M15
21	Weiss (White)	Winter camouflage, identification, codes and tactical markings	M10
22	Schwarz (Black)	Night fighter and bomber finishes, code letters, national markings	M6
23	Rot (Red)	Identity letters, tactical markings	M19
24	Dunkelblau (Dark blue)	Identity letters	4 parts M27 1 part M12
25	Hellgrun (Light green	Identity letters	3 parts G6 1 part G2
26	Braun (Brown)	Early camouflage colour	8 parts M1 1 part M6
27	Gelb (Yellow)	Identity letters	8 parts M15 1 part M1
28	Weinrot (Ruby)	Rarely used colour for identification letters	equal parts M6 and M12
65	Hellblau (Light blue)	Undersurfaces	M25
66	Schwarzgrau (Black grey)	Upper surface colour	equal parts M22 and M21
70	Schwarzgrün (Black green)	Upper surface colour	M17 nearest equivalent
71	Dunkelgrün (Dark green)	Upper surface colour	2 parts M17 1 part M22
72	Grün (Green)	Upper surface colour	2 parts M17 1 part M6
73	Grün (Green)	Upper surface colour	M21 nearest equivalent
74	Dunkelgrau (Dark grey)	Upper surface colour	M22
75	Mittelgrau (Medium grey)	Upper surface colour	M2
76	Hellgrau (Light grey)	Camouflage colour used on top and bottom surfaces and sometimes overall	2 parts M13 1 part M25
79	Sand-gelb (Sand yellow)	Tropical camouflage	2 parts M5, 1 part M15, 1 part M9

It should be noted that where gloss colours have been used they should have a coat of matt varnish applied over them when they are dry. The colours used on Luftwaffe aircraft were not 'dead' flat but had a very slight sheen. This is difficult to reproduce in small scale where it is probably best to leave the model with a matt finish. On larger scales a coat of matt varnish should add just sufficient lustre to give an appearance that closely approximates the original finish. But remember, weather, exposure to the elements, and general service use, soon brought deterioration to any aircraft's finish. It is particularly noteworthy that in tropical climates colours would fade very quickly when exposed to the strong sunlight prevalent in such areas.